THE TACIT DIMENSION

THE TACIT DIMENSION

MICHAEL POLANYI

With a New Foreword
by Amartya Sen

THE UNIVERSITY OF CHICAGO PRESS

CHICAGO AND LONDON

The University of Chicago Press, Chicago 60637
The University of Chicago Press, Ltd., London
© 1966 by Michael Polanyi
Foreword © 2009 by Amartya Sen

Published by arrangement with The Doubleday Broadway
Publishing Group, a division of Random House, Inc.
University of Chicago Press edition, 2009
Printed in the United States of America

18 17 16 15 14 13 12 11   4 5

ISBN-13: 978-0-226-67298-4 (paper)
ISBN-10: 0-226-67298-0 (paper)

Library of Congress Cataloging-in-Publication Data

Polanyi, Michael, 1891–1976.
   The tacit dimension / Michael Polanyi ; with a new foreword
by Amartya Sen.
      p.   cm.
   Includes bibliographical references and index.
   ISBN-13: 978-0-226-67298-4 (pbk. : alk. paper)
   ISBN-10: 0-226-67298-0 (pbk. : alk. paper)   1. Knowledge,
Theory of.   2. Science — Philosophy.   I. Sen, Amartya Kumar.
II. Title.
   BD161.P724 2009
   121 — dc22

                                          2008047128

⊚ The paper used in this publication meets the minimum
requirements of the American National Standard for Information
Sciences — Permanence of Paper for Printed Library Materials,
ANSI Z39.48-1992.

CONTENTS

ACKNOWLEDGMENTS

I am grateful to Yale University for extending to me the invitation to deliver the Terry Lectures of 1962, from which this book has been developed. The theme of the book took shape during my years at Merton College, Oxford, as Senior Research Fellow, and was first expounded in public lectures delivered at the University of Virginia in 1961. I developed these ideas further during my stay at the Center for Advanced Studies in the Behavioral Sciences at Palo Alto, at Duke University where I lectured during the summer of 1964, and at Wesleyan University where I was a Senior Fellow at its Center for Advanced Studies in 1965–66.

I gratefully remember my friends who responded to my thoughts and enriched them. Professor Philip Hallie of Wesleyan, Professor Marjorie Grene of the University of California, and my wife (who also prepared the index) read the manuscript and did much to put it into better shape.

I wish to thank Professor Harry Woolf, Chairman of the Department of the History of Science, for granting permission to use part of my essay "Science and Man's Place in the Universe" which appeared in his *Science as a Cultural Force* (Johns Hopkins Press, 1964).

1

This insightful book by Michael Polanyi was first published in 1966. It is based on his Terry Lectures delivered four years earlier, at Yale University. It is a deeply philosophical book, full of penetrating ideas, particularly about the knowledge of the world and the world of knowledge. The book, like other philosophical works of Polanyi, has received widespread attention and has generated a large literature. In many ways, it has become a part of contemporary culture, and I am delighted that the book is being reissued.

Interestingly enough, philosophy is not, in fact, the subject in which Polanyi achieved his initial fame. In discussing and assessing the contents of this book and its far-reaching implications, it may be helpful to understand the exceptional range of Michael Polanyi's intellectual interests and involvements, since they strongly influenced the nature of the questions that he asked and the kind of answers he presented.

Long before this book was written, Polanyi had become well known across the world as an extraordinarily innovative scientist, particularly in physical chemistry. Polanyi's scientific career had begun early. He published his much-acclaimed first paper, on the chemistry of hydrocephalic liquid, when he was only nineteen, and his frequent contributions continued, through many decades, to receive praise and admiration. In 1933, when Polanyi resigned his academic position at Kaiser-Wilhelm Institute in Germany (in response to the newly emerging Nazi politics) to take up a chair in physical chemistry at Manchester University, his leading role in the world of the natural sciences was already well established—and had been so for many years.

Before going into his investigations in philosophy (I should say, "on the way" to these investigations), Polanyi pursued economics and the social sciences for a number of years. The political issue that engaged him initially was his sense, based on his disillusionment with the Soviet Union, of the tendency toward a "denial of the very existence of scientific thought." But his queries about science in the USSR were supplemented—indeed complemented—by his general interest in the nature of that society and its economy. Two years after moving to his chair of physical chemistry in Manchester, Polanyi wrote a highly critical monograph on Soviet economic practice, and this was followed, five years later, by a more political book, *Contempt of Freedom*. From then on, he wrote a series of other books on a variety of economic, social and political issues, varying from *Patent Reform* (1944) and *Full Employment and Free Trade* (1945), to *Science, Faith and Society* (1946) and *The Logic of Liberty* (1951). There is, obviously, something of a substantial shift here from Polanyi's earlier work in the natural sciences; indeed, he moved in 1948 from his chair in physical chemistry at Manchester University to a new chair of social studies that was specially created for him at the same university.

However, it would be, I think, a mistake to see that "shift" as a "break." Polanyi's experiences and ideas in the natural sciences, along with those in the social sciences, would influence his writings in philosophy in what can be described as the next—the third—stage of Michael Polanyi's intellectual journey, to which this book belongs (as does his earlier book, *Personal Knowledge: Toward a Post-critical Philosophy*, published in 1958). In his extraordinarily ambitious attempt to achieve an understanding of the world—physical as well as mental—through the perspective of knowledge, Polanyi found room for pursuing a huge variety of questions that drew, among other things, on the breadth of his own work in different fields, as well as on his seemingly limitless curiosity about the ideas and analyses presented by scholars in a wide range of fields over thousands of years. Michael Polanyi's polymathic background is, I would argue, quite important in understanding the nature of his philosophical engagement.

2

When Michael Polanyi gave his Terry Lectures in 1962, he had entered the eighth decade of his life. When this book, based on those lectures, was first published in 1966, the *Times Literary Supplement* noted, in an enthusiastic review, "In the rich afterglow of his career as a scientist Dr. Polanyi continues to develop philosophical ideas of great fertility and originality." Polanyi himself begins this book by describing his work on philosophy as "an after-thought to my career as a scientist."

And yet this is a book on science as well. He makes robust use of ideas—his own and those of others—in the physical and social sciences. In giving knowledge such a central position in the comprehension of the world, one has to draw on a deep-rooted understanding of how knowledge emerges and flourishes in the world of nature, especially in the world of human beings and human minds within that larger whole. This is where Polanyi's past work is so important in the genesis of this book. Insights drawn from past scientific works may transcend those works, and yet those works remain critically important for the insights and their use.

The basic insight that launches this book is the understanding that, as Michael Polanyi puts it, "we can know more than we can tell." For example, we can recognize a face clearly enough without being able to tell exactly what features of the face yielded that recognition. The phenomenon relates to Gestalt psychology, which points to the fact that we can integrate the particulars of a physiognomy without being able to identify, in any precise way, those particulars. Polanyi shows that "tacit knowledge" that cannot be easily formalized and put into exact words has a sweeping presence in the world, and he goes on to argue that it is also a central feature of our knowledge of that world.

In the first chapter of the book, Polanyi explores the far-reaching—and sometimes astonishing—implications of tacit knowledge. We get a collection of related propositions here: "Tacit knowledge is shown to account (1) for a valid knowledge of a problem, (2) for the scientist's capacity to pursue it, guided by his sense of approaching his solutions, and (3) for a valid anticipation of the yet indeterminate implications of the discovery arrived at in the end."

3

Polanyi also uses the idea of tacit knowledge to tackle a paradox, discussed by Plato, called "Meno's paradox." This deals with the view that the search for knowledge is an absurdity, since either you know it already, in which case no search is needed, or you do not know what you are looking for, in which case you cannot expect to find it. In contrast, Polanyi argues that if tacit knowledge is a central part of knowledge in general, then we can both (1) know what to look for, and (2) have some idea about what else we may want to know. One implication that Polanyi draws from this perspective is that "the process of formalizing all knowledge to the exclusion of any tacit knowledge is self-defeating." This has subversive implications for the general approach of formalization since it looks for "the kind of lucidity which destroys its subject matter."

Beginning with his exploration of the nature, reach and implications of "tacit knowing," Polanyi goes on to investigate "how the structure of tacit knowing determines the structure of comprehensive entities," and then examines the foundational implications of these recognitions in understanding the nature of the world at different levels. He looks for a framework within which we can define "responsible human action, of which man's moral decisions form but a particular instance." The book ends with a pointer to Polanyi's program of trying "to affiliate our creative endeavors to the organic evolution from which we have arisen," and to understand how all this might relate to "a purpose that bears on eternity." I shall not try to summarize the sophisticated arguments that take Polanyi through this long line of reasoning, but even those who would find the endpoint of the book to be far too ambitious would certainly reap a great many insights from the wide-ranging discussions that Polanyi presents to his readers.

4

A question that may be worth asking relates to Michael Polanyi's continuing status as an "outsider" in professional philosophy, despite his extensive and forceful writings, including this book, on that general subject. Ideas from Michael Polanyi's writings are often cited and used in intellectual discussions, even in professional philosophy, but typically they are treated as suggestions coming from outside the profession.

This is not, in any sense, a failure of Polanyi's purpose. One does not have to be an "insider" in professional philosophy to make powerful philosophical points that engage the attention of a great many people—both inside and outside professional philosophy.

Among the really interesting points to emerge is

the understanding that operations at a higher level cannot be accounted for by the laws governing its particulars forming a lower level. This would militate, for example, against what Polanyi describes as "the predominant view of biologists—that a mechanical explanation of living functions amounts to their explanation in terms of physics and chemistry." The importance of tacit knowledge has implications, Polanyi argues, for the impossibility of depersonalizing knowledge and the difficulty of seeking objectivity in the form of personal detachment. He also uses his line of reasoning to assign a place for authority and to make room for tradition in the enterprise of knowledge. There are a great many interesting points in the book that would engage the reader, even if it is too much to expect that every reader would invariably agree with the conclusions that Polanyi derives from his stimulating lines of inquiry.

5

However, the question does remain: given the reach of his philosophical ideas, why is Michael Polanyi treated as a respected outsider, rather than as an insider, in contemporary Anglo-American philosophy? To say, as some have done, that this is because he was mainly in the tradition of what is called "Continental" philosophy cannot be an adequate answer. For one thing, Anglo-American philosophy does treat Continental philosophers as distinctly professional philosophers, even though working in a different tradition. The Anglo-American philosophers who have no time for, say, Heidegger, would

tend to place his works within professional philosophy—albeit at its misguided end. Also, Michael
Polanyi's work, with his focus on science and ratio-
nality, has much in common with many of the basic ingredients of Anglo-American philosophy.

What then is the answer? It is perhaps relevant to
note in this context that the questions that engaged
Michael Polanyi were not the ones that were central to the debates in professional philosophy at
that time. For example, Polanyi stayed firmly outside both sides of the philosophical divide in the
debates in the declining days of logical positivism.
He explicitly rejected positivist philosophy—not
surprisingly, given his skepticism of detached veri-
fication or falsification as a criterion of soundness
of knowledge. But he also rebuffed the alternatives
that were then emerging. For example, he was not
persuaded by what came to be known as "ordinary
language philosophy," since he did not want to
give that much importance to "linguistic rules"
(Polanyi is more explicit on this subject in his
earlier philosophical book, *Personal Knowledge*).
Even though he spent quite some time in Oxford,
it is hard to see any particular influence of the Oxford analytical school on his work.

On a related matter, Polanyi was in disagreement
with the "early Wittgenstein" of *Tractatus Logico-
Philosophicus*. Among other things, he clearly did
not believe that important recognitions can typically be articulated very "clearly," but nor, unlike
Wittgenstein, did he see that to be an adequate
reason to "be silent" (as advised in the *Tractatus*).
However, the "later Wittgenstein" of *Philosophical
Investigations* did not satisfy Polanyi any more

than did the *Tractatus,* and he expressed himself firmly against relying so much on rules of language. The philosophical divides that engaged Anglo-American philosophers of that period left Polanyi rather cold, and this did not make his integration into mainstream philosophy any easier.

Certainly, being outside these major debates would not have helped Polanyi to become an insider in professional philosophy, but the main reason for his outside status lies, I think, elsewhere. It perhaps lies mostly in the fact that in many ways Polanyi actually *chose* to remain an outsider to mainstream philosophy. His books in philosophy are not written in quite the way standard philosophical books are written. The format of presenting a few sharply defined questions, followed by very detailed—and rather fussy—answers, did not appeal to Michael Polanyi. Nor did he espouse the profession's usual practice of explicitly stating various interpretations of what looks like the same statement and then discussing each interpretation at great length.

Instead, what we get from Polanyi are rather rapid-fire sequences of insights—often deep insights—without much pause for examining alternative interpretations and possible counterarguments. But what may be missed here by professional Anglo-American philosophers can be a source of relief and delight for the general reader interested in philosophical issues (not many general readers, I would suspect, long for a 200-page dissection of the meaning and content of, say, "intentions"). If this is right, then the popularity of Polanyi's work among the general public, on the one hand, and the somewhat distant treatment, on the other, it receives from professional philosophers are not caused by different factors—they relate, to a great extent, to the same features of Polanyi's philosophical writings.

I cannot help feeling that Michael Polanyi would have been quite happy with this trade-off, since he wanted to communicate widely his far-reaching ideas, and also had relatively little patience for finicky discussions in professional philosophy. He would have been content, I think, to be seen as someone with foundational philosophical ideas that actually engaged others, whether inside or outside the narrowly defined boundaries of professional philosophy. The distance from professional mainstream philosophy was, I think, Polanyi's own decision—perhaps a tacit decision, rather than an explicit choice.

The reissuing of this remarkable book gives us a new opportunity to see how far-reaching—and foundational—Michael Polanyi's ideas are, on some of the age-old questions in philosophy. It is a wonderful privilege for me to have a little role in the republication of this short but truly grand contribution to our knowledge and understanding of the world in which we live.

Amartya Sen

This book is an interim report on an inquiry started more than twenty years ago. My ideas were first given a systematic form in *Science, Faith and Society* in 1946. I considered science there as a variant of sensory perception and developed this view into three lectures on the subjects Science and Reality, Authority and Conscience, and Dedication or Servitude. In my Gifford Lectures (Aberdeen, 1951–52) I greatly expanded these themes by including the whole range of knowledge rooted in the life of animals and men. The result was *Personal Knowledge* (1958), supplemented by a theory of historiography in a small book, *The Study of Man* (1959). Since then I have continued this inquiry and published some twenty essays (listed in the Related Bibliography) as well as piled up much unpublished writing.

The present volume is the first account in book form of the work done during these nine years. The delay was caused by hope and by fear. The lure of the next bend behind which new sights might appear distracts us from the labor of taking stock, and the effect of this distraction is reinforced by the anxiety that our theories might be defeated at the next turn.

It took me three years to feel assured that my reply to the *Meno* in the Terry Lectures was right. This has at last been cleared up to my satisfaction in my essay "The Creative Imagination," published in *Chemical Engineering News* (Vol. 44 [1966], No. 17).* It ap-

* This essay was written for the Study Group on Foundations of Cultural Unity held at Bowdoin College in August 1965, and will also be published in their pro-

pears now also that what I have said in the Terry Lectures about our capacity for seeing and pursuing problems had been said long ago in *Science, Faith and Society*. Besides, my hesitant suggestion in the Terry Lectures that tacit knowing is the way in which we are aware of neural processes in terms of perceived objects has been consolidated in my essay "The Structure of Consciousness," recently published in *Brain* (Vol. 88 [1965], Part IV, pp. 799–810).

The Terry Lectures of 1962 thus give a correct summary of my position. The text of Lectures 1 and 2 has been retained virtually unchanged. The opening and closing sections of Lecture 3 are essentially retained, but the link between them has been reshaped by insertion of a more detailed account of the pursuit of science in society.

Viewing the content of these pages from the position reached in *Personal Knowledge* and *The Study of Man* eight years ago, I see that my reliance on the necessity of commitment has been reduced by working out the structure of tacit knowing. This structure shows that all thought contains components of which we are subsidiarily aware in the focal content of our thinking, and that all thought dwells in its subsidiaries, as if they were parts of our body. Hence thinking is not only necessarily intentional, as Brentano has taught: it is also necessarily fraught with the roots that it embodies. It has a *from-to* structure.

A variety of operations based on this structure has proved it to be a richly revealing representation of thought. The fact that it is impossible to account for the nature and justification of knowledge by a series of strictly explicit operations appears obvious in its light, without invoking deeper forms of commitment. And

ceedings, "Toward a Unity of Knowledge" (*Psychological Issues*, in press).

something else also comes into view, which bears on the counterpole of explicit thought in existentialism. Since subsidiaries are used as we use our body, all novel thought is seen to be an existential commitment.

We have then a handy model on which we can reproduce all major existential actions without touching on the great issues of man's fate. I shall show, for example, that when originality breeds new values, it breeds them tacitly, by implication; we cannot choose explicitly a set of new values, but must submit to them by the very act of creating or adopting them.

I have shown that any attempt to avoid the responsibility for shaping the beliefs which we accept as true is absurd; but the existentialist claim of choosing our beliefs from zero is now proved absurd too. Thought can live only on grounds which we adopt in the service of a reality to which we submit.

Center for Advanced Studies
Wesleyan University
April 1966

—1—

TACIT KNOWING

SOME OF you may know that I turned to philosophy as an afterthought to my career as a scientist. I would like to tell you what I was after in making this change, for it will also explain the general task to which my present lecture should introduce us.

I first met questions of philosophy when I came up against the Soviet ideology under Stalin which denied justification to the pursuit of science. I remember a conversation I had with Bukharin in Moscow in 1935. Though he was heading toward his fall and execution three years later, he was still a leading theoretician of the Communist party. When I asked him about the pursuit of pure science in Soviet Russia, he said that pure science was a morbid symptom of a class society; under socialism the conception of science pursued for its own sake would disappear, for the interests of scientists would spontaneously turn to problems of the current Five-Year Plan.

I was struck by the fact that this denial of the very existence of independent scientific thought came from a socialist theory which derived its tremendous persuasive power from its claim to scientific certainty. The scientific outlook appeared to have produced a mechanical conception of man and history in which there was no place for science itself. This conception denied altogether any intrinsic

3

power to thought and thus denied also any grounds for claiming freedom of thought.

I saw also that this self-immolation of the mind was actuated by powerful moral motives. The mechanical course of history was to bring universal justice. Scientific skepticism would trust only material necessity for achieving universal brotherhood. Skepticism and utopianism had thus fused into a new skeptical fanaticism.

It seemed to me then that our whole civilization was pervaded by the dissonance of an extreme critical lucidity and an intense moral conscience, and that this combination had generated both our tight-lipped modern revolutions and the tormented self-doubt of modern man outside revolutionary movements. So I resolved to inquire into the roots of this condition.

My search has led me to a novel idea of human knowledge from which a harmonious view of thought and existence, rooted in the universe, seems to emerge.

I shall reconsider human knowledge by starting from the fact that *we can know more than we can tell*. This fact seems obvious enough; but it is not easy to say exactly what it means. Take an example. We know a person's face, and can recognize it among a thousand, indeed among a million. Yet we usually cannot tell how we recognize a face we know. So most of this knowledge cannot be put into words. But the police have recently introduced a method by which we can communicate much of this knowledge. They have made a large collection of pictures showing a variety of noses, mouths, and other features. From these the witness selects the

4

particulars of the face he knows, and the pieces can then be put together to form a reasonably good likeness of the face. This may suggest that we can communicate, after all, our knowledge of a physiognomy, provided we are given adequate means for expressing ourselves. But the application of the police method does not change the fact that previous to it we did know more than we could tell at the time. Moreover, we can use the police method only by knowing how to match the features we remember with those in the collection, and we cannot tell how we do this. This very act of communication displays a knowledge that we cannot tell.

There are many other instances of the recognition of a characteristic physiognomy—some commonplace, others more technical—which have the same structure as the identification of a person. We recognize the moods of the human face, without being able to tell, except quite vaguely, by what signs we know it. At the universities great efforts are spent in practical classes to teach students to identify cases of diseases and specimens of rocks, of plants and animals. All descriptive sciences study physiognomies that cannot be fully described in words, nor even by pictures.

But can it not be argued, once more, that the possibility of teaching these appearances by practical exercises proves that we can tell our knowledge of them? The answer is that we can do so only by relying on the pupil's intelligent co-operation for catching the meaning of the demonstration. Indeed, any definition of a word denoting an external thing must ultimately rely on pointing at such a thing. This naming-cum-pointing is called "an ostensive

5

definition"; and this philosophic expression conceals a gap to be bridged by an intelligent effort on the part of the person to whom we want to tell what the word means. Our message had left something behind that we could not tell, and its reception must rely on it that the person addressed will discover that which we have not been able to communicate.

Gestalt psychology has demonstrated that we may know a physiognomy by integrating our awareness of its particulars without being able to identify these particulars, and my analysis of knowledge is closely linked to this discovery of Gestalt psychology. But I shall attend to aspects of Gestalt which have been hitherto neglected. Gestalt psychology has assumed that perception of a physiognomy takes place through the spontaneous equilibration of its particulars impressed on the retina or on the brain. However, I am looking at Gestalt, on the contrary, as the outcome of an active shaping of experience performed in the pursuit of knowledge. This shaping or integrating I hold to be the great and indispensable tacit power by which all knowledge is discovered and, once discovered, is held to be true.

The structure of Gestalt is then recast into a logic of tacit thought, and this changes the range and perspective of the whole subject. The highest forms of integration loom largest now. These are manifested in the tacit power of scientific and artistic genius. The art of the expert diagnostician may be listed next, as a somewhat impoverished form of discovery, and we may put in the same class the performance of skills, whether artistic, athletic, or technical. We have here examples of knowing, both

6

of a more intellectual and more practical kind; both the *"wissen"* and *"können"* of the Germans, or the "knowing what" and the "knowing how" of Gilbert Ryle. These two aspects of knowing have a similar structure and neither is ever present without the other. This is particularly clear in the art of diagnosing, which intimately combines skillful testing with expert observation. I shall always speak of "knowing," therefore, to cover both practical and theoretical knowledge. We can, accordingly, interpret the use of tools, of probes, and of pointers as further instances of the art of knowing, and may add to our list also the denotative use of language, as a kind of verbal pointing.

Perception, on which Gestalt psychology centered its attention, now appears as the most impoverished form of tacit knowing. As such it will be shown to form the bridge between the higher creative powers of man and the bodily processes which are prominent in the operations of perception.

Some recent psychological experiments have shown in isolation the principal mechanism by which knowledge is tacitly acquired. Many of you have heard of these experiments as revealing the diabolical machinery of hidden persuasion. Actually, they are but elementary demonstrations of the faculty by which we apprehend the relation between two events, both of which we know, but only one of which we can tell.

Following the example set by Lazarus and Mc-Cleary in 1949, psychologists call the exercise of this faculty a process of "subception."[1] These authors presented a person with a large number of nonsense syllables, and after showing certain of the syllables,

7

they administered an electric shock. Presently the person showed symptoms of anticipating the shock at the sight of "shock syllables"; yet, on questioning, he could not identify them. He had come to know when to expect a shock, but he could not tell what made him expect it. He had acquired a knowledge similar to that which we have when we know a person by signs which we cannot tell.

Another variant of this phenomenon was demonstrated by Eriksen and Kuethe in 1958.[2] They exposed a person to a shock whenever he happened to utter associations to certain "shock words." Presently, the person learned to forestall the shock by avoiding the utterance of such associations, but, on questioning, it appeared that he did not know he was doing this. Here the subject got to know a practical operation, but could not tell how he worked it. This kind of subception has the structure of a skill, for a skill combines elementary muscular acts which are not identifiable, according to relations that we cannot define.

These experiments show most clearly what is meant by saying that one can know more than one can tell. For the experimental arrangement wards off the suspicion of self-contradiction, which is not easy to dispel when anyone speaks of things he knows and cannot tell. This is prevented here by the division of roles between the subject and the observer. The experimenter observes that another person has a certain knowledge that he cannot tell, and so no one speaks of a knowledge he himself has and cannot tell.

We may carry forward, then, the following result. In both experiments that I have cited, subception was induced by electric shock. In the first series the subject was shocked after being shown certain non-sense syllables, and he learned to expect this event. In the second series he learned to suppress the uttering of certain associations, which would evoke the shock. In both cases the shock-producing particulars remained tacit. The subject could not identify them, yet he relied on his awareness of them for anticipating the electric shock.

Here we see the basic structure of tacit knowing. It always involves two things, or two kinds of things. We may call them the two terms of tacit knowing. In the experiments the shock syllables and shock associations formed the first term, and the electric shock which followed them was the second term. After the subject had learned to connect these two terms, the sight of the shock syllables evoked the expectation of a shock and the utterance of the shock associations was suppressed in order to avoid shock. Why did this connection remain tacit? It would seem that this was due to the fact that the subject was riveting his attention on the electric shock. He was relying on his awareness of the shock-producing particulars only in their bearing on the electric shock. We may say that he learned to rely on his awareness of these particulars for the purpose of attending to the electric shock.

Here we have the basic definition of the logical relation between the first and second term of a tacit knowledge. It combines two kinds of knowing. We know the electric shock, forming the second term, by attending to it, and hence the subject is *specifiably* known. But we know the shock-producing particulars only by relying on our own awareness of

9

them for attending to something else, namely the electric shock, and hence our knowledge of them remains *tacit*. This is how we come to know these particulars, without becoming able to identify them. Such is the *functional relation* between the two terms of tacit knowing: *we know the first term only by relying on our awareness of it for attending to the second.*

In his book on freedom of the will, Austin Farrar has spoken at one point of *disattending from* certain things for attending *to* others. I shall adopt a variant of this usage by saying that in an act of tacit knowing we *attend from* something for attending *to* something else; namely, *from* the first term *to* the second term of the tacit relation. In many ways the first term of this relation will prove to be nearer to us, the second further away from us. Using the language of anatomy, we may call the first term *proximal*, and the second term *distal*. It is the proximal term, then, of which we have a knowledge that we may not be able to tell.

In the case of a human physiognomy, I would now say that we rely on our awareness of its features for attending to the characteristic appearance of a face. We are attending *from* the features *to* the face, and thus may be unable to specify the features. And I would say, likewise, that we are relying on our awareness of a combination of muscular acts for attending to the performance of a skill. We are attending *from* these elementary movements *to* the achievement of their joint purpose, and hence are usually unable to specify these elementary acts. We may call this the *functional structure* of tacit knowing.

10

But we may ask: does not the *appearance* of the experimental setting—composed of the nonsense syllables and the electric shocks—undergo some change when we learn to anticipate a shock at the sight of certain syllables? It does, and in a very subtle way. The expectation of a shock, which at first had been vague and unceasing, now becomes sharply fluctuating; it suddenly rises at some moments and drops between them. So we may say that even though we do not learn to recognize the shock syllables as distinct from other syllables, we do become aware of facing a shock syllable in terms of the apprehension it evokes in us. In other words, we are aware of seeing these syllables in terms of that on which we are focusing our attention, which is the probability of an electric shock. Applying this to the case of a physiognomy, we may say that we are aware of its features in terms of the physiognomy to which we are attending. In the exercise of a skill, we are aware of its several muscular moves in terms of the performance to which our attention is directed. We may say, in general, that we are aware of the proximal term of an act of tacit knowing in the appearance of its distal term; we are aware of that *from* which we are attending *to* another thing, in the *appearance* of that thing. We may call this the *phenomenal structure* of tacit knowing.

But there is a significance in the relation of the two terms of tacit knowing which combines its functional and phenomenal aspects. When the sight of certain syllables makes us expect an electric shock, we may say that they *signify* the approach of a shock. This is their *meaning* to us. We could

11

say, therefore, that when shock syllables arouse an apprehension in us, without our being able to identify the syllables which arouse it, we know these syllables only in terms of their meaning. It is their meaning to which our attention is directed. It is in terms of their meaning that they enter into the appearance of that *to* which we are attending *from* them.

We could say, in this sense, that a characteristic physiognomy is the meaning of its features; which is, in fact, what we do say when a physiognomy expresses a particular mood. To identify a physiognomy would then amount to relying on our awareness of its features for attending to their joint meaning. This may sound far-fetched, because the meaning of the features is observed at the same spot where the features are situated, and hence it is difficult to separate mentally the features from their meaning. Yet, the fact remains that the two are distinct, since we may know a physiognomy without being able to specify its particulars.

To see more clearly the separation of a meaning from that which has this meaning, we may take the example of the use of a probe to explore a cavern, or the way a blind man feels his way by tapping with a stick. For here the separation of the two is wide, and we can also observe here the process by which this separation gradually takes place. Anyone using a probe for the first time will feel its impact against his fingers and palm. But as we learn to use a probe, or to use a stick for feeling our way, our awareness of its impact on our hand is transformed into a sense of its point touching the objects we are exploring. This is how an interpretative effort

12

transposes meaningless feelings into meaningful ones, and places these at some distance from the original feeling. We become aware of the feelings in our hand in terms of their meaning located at the tip of the probe or stick to which we are attending. This is so also when we use a tool. We are attending to the meaning of its impact on our hands in terms of its effect on the things to which we are applying it. We may call this the *semantic aspect* of tacit knowing. All meaning tends to be displaced *away from ourselves*, and that is in fact my justification for using the terms "proximal" and "distal" to describe the first and second terms of tacit knowing.

From the three aspects of tacit knowing that I have defined so far—the functional, the phenomenal, and the semantic—we can deduce a fourth aspect, which tells us what tacit knowing is a knowledge of. This will represent its *ontological* aspect. Since tacit knowing establishes a meaningful relation between two terms, we may identify it with the *understanding* of the comprehensive entity which these two terms jointly constitute. Thus the proximal term represents the *particulars* of this entity, and we can say, accordingly, that we comprehend the entity by relying on our awareness of its particulars for attending to their joint meaning.

This analysis can be applied with interesting results to the case of visual perception. Physiologists long ago established that the way we see an object is determined by our awareness of certain efforts inside our body, efforts which we cannot feel in themselves. We are aware of these things going on inside our body in terms of the position, size, shape, and motion of an object, to which we are attending.

In other words we are attending *from* these internal processes *to* the qualities of things outside. These qualities are what those internal processes *mean* to us. The transposition of bodily experiences into the perception of things outside may now appear, therefore, as an instance of the transposition of meaning away from us, which we have found to be present to some extent in all tacit knowing.

But it may be said that the feelings transposed by perception differ from those transposed by the use of tools or probes, by being hardly noticeable in themselves previous to their transposition. An answer to this—or at least part of an answer to it—is to be found in experiments extending subception to subliminal stimuli. Hefferline and collaborators have observed that when spontaneous muscular twitches, unfelt by the subject—but observable externally by a million-fold amplification of their action currents—were followed by the cessation of an unpleasant noise, the subject responded by increasing the frequency of the twitches and thus silencing the noise much of the time.[8] Tacit knowing is seen to operate here on an internal action that we are quite incapable of controlling or even feeling in itself. We become aware of our operation of it only in the silencing of a noise. This experimental result seems closely analogous to the process by which we become aware of subliminal processes inside our body in the perception of objects outside.

This view of perception, that it is an instance of the transposition of feelings which we found in the use of probes and in the process of subception, is borne out by the fact that the capacity to see external objects must be acquired, like the use of

14

probes and the feats of subception, by a process of learning which can be laborious.

Modern philosophers have argued that perception does not involve projection, since we are not previously aware of the internal processes which we are supposed to have projected into the qualities of things perceived. But we have now established that projection of this very kind is present in various instances of tacit knowing. Moreover, the fact that we do not originally sense the internal processes in themselves now appears irrelevant. We may venture, therefore, to extend the scope of tacit knowing to include neural traces in the cortex of the nervous system. This would place events going on inside our brain on the same footing as the subliminal twitches operated by Hefferline's subjects.*

This brings us to the point at which I hinted when I first mentioned perception as an instance of tacit knowing. I said that by elucidating the way our bodily processes participate in our perceptions we will throw light on the bodily roots of all thought, including man's highest creative powers. Let me show this now.

Our body is the ultimate instrument of all our external knowledge, whether intellectual or practical. In all our waking moments we are *relying* on

---

* Such a hypothesis does not explain how perceived sights, or any other state of consciousness, arise in conjunction with neural processes. It merely applies the principle that wherever some process in our body gives rise to consciousness in us, our tacit knowing of the process will make sense of it in terms of an experience to which we are attending.

our awareness of contacts of our body with things outside for *attending* to these things. Our own body is the only thing in the world which we normally never experience as an object, but experience always in terms of the world to which we are attending from our body. It is by making this intelligent use of our body that we feel it to be our body, and not a thing outside.

I have described how we learn to feel the end of a tool or a probe hitting things outside. We may regard this as the transformation of the tool or probe into a sentient extension of our body, as Samuel Butler has said. But our awareness of our body for attending to things outside it suggests a wider generalization of the feeling we have of our body. Whenever we use certain things for attending *from* them to other things, in the way in which we always use our own body, these things change their appearance. They appear to us now in terms of the entities to which we are attending *from* them, just as we feel our own body in terms of the things outside to which we are attending *from* our body. In this sense we can say that when we make a thing function as the proximal term of tacit knowing, we incorporate it in our body—or extend our body to include it—so that we come to dwell in it.

The full range of this generalization can only be hinted at here. Indications of its scope may be seen by recalling that, at the turn of the last century, German thinkers postulated that indwelling, or empathy, is the proper means of knowing man and the humanities. I am referring particularly to Dilthey[4] and Lipps.[5] Dilthey taught that the mind of a person can be understood only by reliving its workings;

16

and Lipps represented aesthetic appreciation as an entering into a work of art and thus dwelling in the mind of its creator. I think that Dilthey and Lipps described here a striking form of tacit knowing as applied to the understanding of man and of works of art, and that they were right in saying that this could be achieved only by indwelling. But my analysis of tacit knowing shows that they were mistaken in asserting that this sharply distinguished the humanities from the natural sciences. Indwelling, as derived from the structure of tacit knowing, is a far more precisely defined act than is empathy, and it underlies all observations, including all those described previously as indwelling.

We meet with another indication of the wide functions of indwelling when we find acceptance to moral teachings described as their *interiorization*. To interiorize is to identify ourselves with the teachings in question, by making them function as the proximal term of a tacit moral knowledge, as applied in practice. This establishes the tacit framework for our moral acts and judgments. And we can trace this kind of indwelling to logically similar acts in the practice of science. To rely on a theory for understanding nature is to interiorize it. For we are attending from the theory to things seen in its light, and are aware of the theory, while thus using it, in terms of the spectacle that it serves to explain. This is why mathematical theory can be learned only by practicing its application: its true knowledge lies in our ability to use it.

The identification of tacit knowing with indwelling involves a shift of emphasis in our conception of tacit knowing. We had envisaged tacit knowing

17

in the first place as a way to know more than we can tell. We identified the two terms of tacit knowing, the proximal and the distal, and recognized the way we attend *from* the first *to* the second, thus achieving an integration of particulars to a coherent entity to which we are attending. Since we were not attending to the particulars in themselves, we could not identify them: but if we now regard the integration of particulars as an interiorization, it takes on a more positive character. It now becomes a means of making certain things function as the proximal terms of tacit knowing, so that instead of observing them in themselves, we may be aware of them in their bearing on the comprehensive entity which they constitute. It brings home to us that it is not by looking at things, but by dwelling in them, that we understand their joint meaning.

We can see now how an unbridled lucidity can destroy our understanding of complex matters. Scrutinize closely the particulars of a comprehensive entity and their meaning is effaced, our conception of the entity is destroyed. Such cases are well known. Repeat a word several times, attending carefully to the motion of your tongue and lips, and to the sound you make, and soon the word will sound hollow and eventually lose its meaning. By concentrating attention on his fingers, a pianist can temporarily paralyze his movement. We can make ourselves lose sight of a pattern or physiognomy by examining its several parts under sufficient magnification.

Admittedly, the destruction can be made good by interiorizing the particulars once more. The word uttered again in its proper context, the pianist's fingers used again with his mind on his music, the features of a physiognomy and the details of a pattern glanced at once more from a distance: they all come to life and recover their meaning and their comprehensive relationship.

But it is important to note that this recovery never brings back the original meaning. It may improve on it. Motion studies, which tend to paralyze a skill, will improve it when followed by practice. The meticulous dismembering of a text, which can kill its appreciation, can also supply material for a much deeper understanding of it. In these cases, the detailing of particulars, which by itself would destroy meaning, serves as a guide to their subsequent integration and thus establishes a more secure and more accurate meaning of them.

But the damage done by the specification of particulars may be irremediable. Meticulous detailing may obscure beyond recall a subject like history, literature, or philosophy. Speaking more generally, the belief that, since particulars are more tangible, their knowledge offers a true conception of things is fundamentally mistaken.

Of course, tacit reintegration of particulars is not the only way to recover their meaning, destroyed by focusing our attention on them. The destructive analysis of a comprehensive entity can be counteracted in many cases by explicitly stating the relation between its particulars. Where such explicit integration is feasible, it goes far beyond the range of tacit integration. Take the case of a machine. One can learn to use it skillfully, without knowing exactly how it works. But the engineer's understanding of its construction and operation goes much deeper.

We possess a practical knowledge of our own body, but the physiologist's theoretical knowledge of it is far more revealing. The formal rules of prosody may deepen our understanding of so delicate a thing as a poem.

But my examples show clearly that, in general, an explicit integration cannot replace its tacit counterpart. The skill of a driver cannot be replaced by a thorough schooling in the theory of the motorcar; the knowledge I have of my own body differs altogether from the knowledge of its physiology; and the rules of rhyming and prosody do not tell me what a poem told me, without any knowledge of its rules.

We are approaching here a crucial question. The declared aim of modern science is to establish a strictly detached, objective knowledge. Any falling short of this ideal is accepted only as a temporary imperfection, which we must aim at eliminating. But suppose that tacit thought forms an indispensable part of all knowledge, then the ideal of eliminating all personal elements of knowledge would, in effect, aim at the destruction of all knowledge. The ideal of exact science would turn out to be fundamentally misleading and possibly a source of devastating fallacies.

I think I can show that the process of formalizing all knowledge to the exclusion of any tacit knowing is self-defeating. For, in order that we may formalize the relations that constitute a comprehensive entity, for example, the relations that constitute a frog, this entity, i.e., the frog, must be first identified informally by tacit knowing; and, indeed, the meaning of a mathematical theory of the frog lies in its

20

continued bearing on this still tacitly known frog. Moreover, the act of bringing a mathematical theory to bear on its subject is itself a tacit integration of the kind we have recognized in the use of a denotative word for designating its object. And we have seen also that a true knowledge of a theory can be established only after it has been interiorized and extensively used to interpret experience. Therefore: a mathematical theory can be constructed only by relying on *prior* tacit knowing and can function as a theory only *within* an act of tacit knowing, which consists in our attending *from* it to the previously established experience on which it bears. Thus the ideal of a comprehensive mathematical theory of experience which would eliminate all tacit knowing is proved to be self-contradictory and logically unsound.

But I must not rest my case on such an abstract argument. Let me finish this lecture, therefore, by presenting you with a most striking concrete example of an experience that cannot possibly be represented by any exact theory. It is an experience within science itself: the experience of seeing a problem, as a scientist sees it in his pursuit of discovery.

It is a commonplace that all research must start from a problem. Research can be successful only if the problem is good; it can be original only if the problem is original. But how can one see a problem, any problem, let alone a good and original problem? For to see a problem is to see something that is hidden. It is to have an intimation of the coherence of hitherto not comprehended particulars. The problem is good if this intimation is true; it is original if

21

no one else can see the possibilities of the comprehension that we are anticipating. To see a problem that will lead to a great discovery is not just to see something hidden, but to see something of which the rest of humanity cannot have even an inkling. All this is a commonplace; we take it for granted, without noticing the clash of self-contradiction entailed in it. Yet Plato has pointed out this contradiction in the *Meno*. He says that to search for the solution of a problem is an absurdity; for either you know what you are looking for, and then there is no problem; or you do not know what you are looking for, and then you cannot expect to find anything.

The solution which Plato offered for this paradox was that all discovery is a remembering of past lives. This explanation has hardly ever been accepted, but neither has any other solution been offered for avoiding the contradiction. So we are faced with the fact that, for two thousand years and more, humanity has progressed through the efforts of people solving difficult problems, while all the time it could be shown that to do this was either meaningless or impossible. We have here the classical case of Poe's *Purloined Letter*, of the momentous document lying casually in front of everybody, and hence overlooked by all. For the *Meno* shows conclusively that if all knowledge is explicit, i.e., capable of being clearly stated, then we cannot know a problem or look for its solution. And the *Meno* also shows, therefore, that if problems nevertheless exist, and discoveries can be made by solving them, we can know things, and important things, that we cannot tell.

The kind of tacit knowledge that solves the paradox of the *Meno* consists in the intimation of something hidden, which we may yet discover. There exists another important manifestation of these mental powers. We are often told that great scientific discoveries are marked by their fruitfulness; and this is true. But how can we recognize truth by its fruitfulness? Can we recognize that a statement is true by appreciating the wealth of its yet undiscovered consequences? This would of course be nonsensical, if we had to know explicitly what was yet undiscovered. But it makes sense if we admit that we can have a tacit foreknowledge of yet undiscovered things. This is indeed the kind of foreknowledge the Copernicans must have meant to affirm when they passionately maintained, against heavy pressure, during one hundred and forty years before Newton proved the point, that the heliocentric theory was not merely a convenient way of computing the paths of planets, but was really true.

It appears, then, that to know that a statement is true is to know more than we can tell and that hence, when a discovery solves a problem, it is itself fraught with further intimations of an indeterminate range, and that furthermore, when we accept the discovery as true, we commit ourselves to a belief in all these as yet undisclosed, perhaps as yet unthinkable, consequences.

Since we have no explicit knowledge of these unknown things, there can also be no explicit justification of a scientific truth. But as we can know a problem, and feel sure that it is pointing to something hidden behind it, we can be aware also of the hidden implications of a scientific discovery, and feel confident that they will prove right. We feel sure of this, because in contemplating the discovery we are looking at it not only in itself but, more significantly, as a clue to a reality of which it is a manifestation. The pursuit of discovery is conducted from the start in these terms; all the time we are guided by sensing the presence of a hidden reality toward which our clues are pointing; and the discovery which terminates and satisfies this pursuit is still sustained by the same vision. It claims to have made contact with reality: a reality which, being real, may yet reveal itself to future eyes in an indefinite range of unexpected manifestations.

We have here reached our main conclusions. Tacit knowing is shown to account (1) for a valid knowledge of a problem, (2) for the scientist's capacity to pursue it, guided by his sense of approaching its solution, and (3) for a valid anticipation of the yet indeterminate implications of the discovery arrived at in the end.

Such indeterminate commitments are necessarily involved in any act of knowing based on indwelling. For such an act relies on interiorizing particulars to which we are not attending and which, therefore, we may not be able to specify, and relies further on our attending from these unspecifiable particulars to a comprehensive entity connecting them in a way we cannot define. This kind of knowing solves the paradox of the *Meno* by making it possible for us to know something so indeterminate as a problem or a hunch, but when the use of this faculty turns out to be an indispensable element of all knowing, we are forced to conclude that all knowledge is of the same kind as the knowledge of a problem.

This is in fact our result. We must conclude that

the paradigmatic case of scientific knowledge, in which all the faculties that are necessary for finding and holding scientific knowledge are fully developed, is the knowledge of an approaching discovery.

To hold such knowledge is an act deeply committed to the conviction that there is something there to be discovered. It is personal, in the sense of involving the personality of him who holds it, and also in the sense of being, as a rule, solitary; but there is no trace in it of self-indulgence. The discoverer is filled with a compelling sense of responsibility for the pursuit of a hidden truth, which demands his services for revealing it. His act of knowing exercises a personal judgment in relating evidence to an external reality, an aspect of which he is seeking to apprehend.

The anticipation of discovery, like discovery itself, may turn out to be a delusion. But it is futile to seek for strictly impersonal criteria of its validity, as positivistic philosophies of science have been trying to do for the past eighty years or so. To accept the pursuit of science as a reasonable and successful enterprise is to share the kind of commitments on which scientists enter by undertaking this enterprise. You cannot formalize the act of commitment, for you cannot express your commitment non-committally. To attempt this is to exercise the kind of lucidity which destroys its subject matter. Hence the failure of the positivist movement in the philosophy of science. The difficulty is to find a stable alternative to its ideal of objectivity. This is indeed the task for which the theory of tacit knowing should prepare us.

—2—
EMERGENCE

I HAVE GIVEN you an account of the way we exercise our tacit powers of knowing. The things that we know in this way included problems and hunches, physiognomies and skills, the use of tools, probes, and denotative language, and my list extended all the way to include the primitive knowledge of external objects perceived by our senses. Indeed, the structure of perception throws light on all the rest. Because our body is involved in the perception of objects, it participates thereby in our knowing of all other things outside. Moreover, we keep expanding our body into the world, by assimilating to it sets of particulars which we integrate into reasonable entities. Thus do we form, intellectually and practically, an interpreted universe populated by entities, the particulars of which we have interiorized for the sake of comprehending their meaning in the shape of coherent entities.

Consider the situation where two persons share the knowledge of the same comprehensive entity—of an entity which one of them produces and the other apprehends. Such is the case when one person has formed a message and the other has received it. But the characteristic features of the situation are seen more clearly if we consider the way one man comes to understand the skillful performance of another man. He must try to combine mentally the movements which the performer combines practically and he must combine them in a pattern similar

29

to the performer's pattern of movements. Two kinds of indwelling meet here. The performer co-ordinates his moves by dwelling in them as parts of his body, while the watcher tries to correlate these moves by seeking to dwell in them from outside. He dwells in these moves by interiorizing them. By such exploratory indwelling the pupil gets the feel of a master's skill and may learn to rival him.

Nor is this structural kinship between subject and object, and the indwelling of one in the other, present only in the study of a bodily performance. Chess players enter into a master's spirit by rehearsing the games he played, to discover what he had in mind.

Moreover, in both these instances of our entry into the particulars of a comprehensive entity, we meet something that accounts for the coherence of the entity. In one case we meet a person skillfully using his body and, in the other, a person cleverly using his mind.

The recognition of a *person* in the performance of a skill or in the conduct of a game of chess is intrinsic to the understanding of these matters. We must surmise that we are faced with some co-ordinated performance, before we can even try to understand it, and must go on trying to pick out the features that are essential to the performance, with a view to the action felt to be at work in it. Hence, the question much discussed by philosophers of how we can infer the existence of other minds from observing their external workings does not arise, for we never do observe these workings in themselves. Indeed, many of them we could not identify, even after we had successfully integrated them with our knowledge of a personal performance, any more than its performer could tell us, except quite vaguely, what the particulars are that his performance co-ordinates.

This is not to say that we gain an understanding of the mind without a process of inquiry. But the inquiry consists, like a scientific inquiry, in picking out clues as such, that is, with a presumed bearing on the presence of something they appear to indicate. And as in a scientific inquiry, many of the clues used will remain unspecifiable and may indeed be subliminal. Such is the effort by which we enter into the intimate structure of a skill or of a game of chess and get to know the powers of the person behind it. This is also the method by which a historian explores a historical personality.

The structural kinship between knowing a mind and pursuing a scientific inquiry throws light on some further points obscured by the false assumption that we start acquiring the knowledge of a mind by observing the workings of the mind in themselves. It indicates that the mind is unsubstantial only in the sense in which a problem is unsubstantial. Indeed, a great mind is an inexhaustible and rewarding problem to the historian and literary scholar, and every person is of infinite concern to one who cares for him. But neither problems nor minds should on this account be set far apart from other things. For an inanimate solid object, too, is known by understanding its particulars, *from which* we attend *to it* as an object.

This brings up a question similar to the one I have just spoken of in respect to our knowledge of other minds. The question is how we infer the existence of a permanent object, from observing its sensible qualities. Some philosophers would dispose of this problem by denying that we ever see anything but objects. But this is not true. We do see the several parts of a camouflaged object as mere patches, and can break down this deception only by an effort to see these fragments meaningfully as an object. These philosophers are right in pointing out that no process of inference takes place either in getting to know a mind, or in seeing a cobblestone, and that it is fruitless, therefore, to inquire about the way such an inference is conducted; but this still leaves us with the task of understanding how we got to know such things tacitly, as we actually do.

The examples which I have mentioned point at a new aspect of this problem of philosophy. The structural kinship between knowing a person and discovering a problem, and the alignment of both with our knowing of a cobblestone, call attention to the greater depth of a person and a problem, as compared with the lesser profundity of a cobblestone. Persons and problems are felt to be more profound, because we expect them yet to reveal themselves in unexpected ways in the future, while cobblestones evoke no such expectation. This capacity of a thing to reveal itself in unexpected ways in the future I attribute to the fact that the thing observed is an aspect of a reality, possessing a significance that is not exhausted by our conception of any single aspect of it. To trust that a thing we know is real is, in this sense, to feel that it has the independence and power for manifesting itself in yet unthought of ways in the future. I shall say, accordingly, that minds and problems possess a deeper reality than

32

cobblestones, although cobblestones are admittedly more real in the sense of being *tangible*. And since I regard the significance of a thing as more important than its tangibility, I shall say that minds and problems are more real than cobblestones. This is to class our knowledge of reality with the kind of foreknowl-·edge which guides scientists to discovery.

With this in mind, we can say now also that man's skillful exercise of his body is a real entity that another person can know, and know only by comprehending it, and that the comprehension of this real entity has the same structure as the entity which is its object.

And we may likewise say—so as to drive the point home—that the skillful conduct of a game of chess by another person is a real entity, knowable by our tacit act of comprehending it, and that this comprehension is similar in structure to that which it comprehends.

You may feel that I have been slow in drawing this conclusion. But I had to make quite sure of it, for it carries far-reaching implications. I said in my previous lecture that the question, what it is that we know by understanding a comprehensive entity, makes an *ontological* reference to it. We have now given a solid content to this hitherto rather vague ontology. We have shown that the kind of comprehensive entities exemplified by skillful human performances are real things; as real as cobblestones and, in view of their far greater independence and power, much *more* real than cobblestones. It seems plausible then to *assume in all other instances of tacit knowing the correspondence between the structure of comprehension and the structure of the*

33

comprehensive entity which is its object. And we would expect then to find the structure of tacit knowing duplicated in the principles which account for the stability and effectiveness of all real comprehensive entities. Let me try to show what this means.

Take two points. (1) Tacit knowing of a coherent entity relies on our awareness of the particulars of the entity for attending to it; and (2) if we switch our attention to the particulars, this function of the particulars is canceled and we lose sight of the entity to which we had attended. The ontological counterpart of this would be (1) that the principles controlling a comprehensive entity would be found to rely for their operations on laws governing the particulars of the entity in themselves; and (2) that at the same time the laws governing the particulars in themselves would never account for the organizing principles of a higher entity which they form. Return to our knowledge of a game of chess and the game itself. The playing of a game of chess is an entity controlled by principles which rely on the observance of the rules of chess; but the principles controlling the game cannot be derived from the rules of chess. The two terms of tacit knowing, the proximal, which includes the particulars, and the distal, which is their comprehensive meaning, would then be seen as two levels of reality, controlled by distinctive principles. The upper one relies for its operations on the laws governing the elements of the lower one in themselves, but these operations of it are not explicable by the laws of the lower level. And we could say that between two such levels a logical relation holds, which corresponds to the fact that the two levels are the two terms of an act of tacit knowing which jointly comprehends them.

I have spoken before of the way we interiorize bits of the universe, and thus populate it with comprehensive entities. The program which I have set out now would change this panorama into a picture of the universe filled with strata of realities, joined together meaningfully in pairs of higher and lower strata.

I could exemplify this by analyzing in these terms the various cases of tacit knowing that I have spoken of already, but I shall rather give new examples, which will take us further by showing pairs of levels which tend to link up into a series forming a hierarchy.

Take the art of making bricks. It relies on its raw materials placed on a level below it. But above the brickmaker there operates the architect, relying on the brickmaker's work, and the architect in his turn has to serve the town planner. To these four successive levels there correspond four successive levels of rules. The laws of physics and chemistry govern the raw material of bricks; technology prescribes the art of brickmaking; architecture teaches the builders; and the rules of town planning control the town planners. I shall discuss in more detail my next example, which is the giving of a speech. It includes five levels; namely the production (1) of voice, (2) of words, (3) of sentences, (4) of style, and (5) of literary composition. Each of these levels is subject to its own laws, as prescribed (1) by phonetics, (2) by lexicography, (3) by grammar, (4) by stylistics, and (5) by literary criticism. These levels form a hierarchy of comprehensive entities, for the principles of each level operate under the control of the next higher level. The voice you produce is shaped into words by a vocabulary; a given vocabulary is shaped into sentences in accordance with grammar; and the sentences can be made to fit into a style, which in its turn is made to convey the ideas of a literary composition. Thus each level is subject to dual control; first, by the laws that apply to its elements in themselves and, second, by the laws that control the comprehensive entity formed by them.[6]

Accordingly, the operations of a higher level cannot be accounted for by the laws governing its particulars forming the lower level. You cannot derive a vocabulary from phonetics; you cannot derive the grammar of a language from its vocabulary; a correct use of grammar does not account for good style; and a good style does not provide the content of a piece of prose. We may conclude then quite generally—in confirmation of what I said when I identified the two terms of tacit knowing with two levels of reality—that it is impossible to represent the organizing principles of a higher level by the laws governing its isolated particulars.

This may seem too obvious to merit such emphasis, but it will prove highly controversial, when I now pass from hierarchies of human skills to the hierarchy of levels found in living beings. The sequence of these levels is built up by the rise of higher forms of life from lower ones. We can see all the levels of evolution at a glance in an individual human being. The most primitive form of life is represented by the growth of the typical human shape, through the process of morphogenesis studied by embryology. Next we have the vegetative functioning of the organism, studied by physiology; and above it there is sentience, rising to perception and to a centrally controlled motoric activity, both of which still belong to the subject of physiology. We rise beyond this at the level of conscious behavior and intellectual action, studied by ethology and psychology; and, uppermost, we meet with man's moral sense, guided by the firmament of his standards.

I shall set aside, for the moment, the question of how far these consecutive levels do form a hierarchy in our sense, and concentrate instead on the fact that all these levels are situated above that of the inanimate, and that hence they all rely for their operations—directly or indirectly—on the laws of physics and chemistry which govern the inanimate. If we then apply the principle that the operations of a higher level can never be derived from the laws governing its isolated particulars, it follows that none of these biotic operations can be accounted for by the laws of physics and chemistry.

Yet it is taken for granted today among biologists that all manifestations of life can ultimately be explained by the laws governing inanimate matter. K. S. Lashley declared this at the Hixon Symposium of 1948, as the common belief of all the participants, without even consulting his distinguished colleagues. Yet this assumption is patent nonsense. The most striking feature of our own existence is our sentience. The laws of physics and chemistry include no conception of sentience, and any system wholly determined by these laws must be insentient. It may be in the interest of science to turn a blind eye on this central fact of the universe, but it is certainly not in the interest of truth. I shall prefer to follow up, on the contrary, the fact that the study of life must ultimately reveal some principles additional to those manifested by inanimate matter, and to prefigure the general outline of one such, yet unknown, principle.

I shall begin this inquiry by a closer scrutiny of the prevailing procedure of modern biologists. While the declared aim of current biology is to explain all the phenomena of life in terms of the laws of physics and chemistry, its actual practice is to attempt an explanation in terms of a *machinery, based on the laws of physics and chemistry*. Biologists think that the substitution of this task for their declared aim is justified, for they assume that a machine based on the laws of physics is explicable by the laws of physics. My first point is that biologists are mistaken in assuming this.

Some authors have pointed out that machines have a purposive character which cannot be derived from the laws of physics and chemistry. But, to obtain the actual relationship of the principles of a machine and the laws governing its parts, we must consider the nature of a machine as a comprehensive entity. This will serve also to consolidate and deepen our conception of the logical structure governing such entities, for in this case it is possible to define with a fair degree of precision the relation by which the parts are integrated to the entity formed by them. I have presented this analysis often elsewhere, and shall therefore state its main points here without arguing them.[7]

Machines are defined by their operational principles, which tell us how the machine works. These operational principles include the definition of the parts composing the machine and give an account of their several functions in the working of the machine; they also state the purpose which the machine is to serve. The machine relies for its functions on certain physical and chemical properties of its parts and on certain physical-chemical processes involved in their joint operation. No more need be required in this respect than that the material of the machine be solid and be ruled by the laws of mechanics.

Engineering and physics are two different sciences. Engineering includes the operational principles of machines and some knowledge of physics bearing on these principles. Physics and chemistry, on the other hand, include no knowledge of the operational principles of machines. Hence a complete physical and chemical topography of an object would not tell us whether it is a machine, and if so, how it works, and for what purpose. Physical and chemical investigations of a machine are meaningless, unless undertaken with a bearing on the previously established operational principles of the machine. But there is an important feature of the machine which its operational principles do not reveal; they can never account for the failure and ultimate breakdown of the machine. And here physics and chemistry effectively come in. Only the physical-chemical structure of a machine can explain its failures. Liability to failure is, as it were, the price paid for embodying operational principles in a material the laws of which ignore these principles. Such a material will eventually cast off the yoke of such foreign control.

But how can a machine which, as an inanimate body, obeys the laws of physics and chemistry fail to be determined by these laws? How can it follow both the laws of nature and its own operational principles as a machine? How does the shaping of inanimate matter in a machine make it capable of success or failure? The answer lies in the word: shaping. Natural laws may mold inanimate matter into distinctive shapes, such as the spheres of the sun and the moon, and into such patterns as that of the solar system. Other shapes can be imposed on matter artificially, and yet without infringing the laws of nature. The operational principles of machines are embodied in matter by such artificial shaping. These principles may be said *to govern the boundary conditions of an inanimate system*—a set of conditions that is explicitly left undetermined by the laws of nature. Engineering provides a determination of such boundary conditions. And this is how an inanimate system can be subject to a dual control on two levels: the operations of the upper level are artificially embodied in the boundaries of the lower level which is relied on to obey the laws of inanimate nature, i.e., physics and chemistry.

We may call the control exercised by the organizational principle of a higher level on the particulars forming its lower level *the principle of marginal control.*

This principle of marginality could be recognized already in the way I described some hierarchies of human performances. You can see, for example, how, in the hierarchy constituting speechmaking,

40

successive working principles control the boundary left indeterminate on the next lower level. Voice production, which is the lowest level of speech, leaves largely open the combination of sounds into words, which is controlled by a vocabulary. Next, a vocabulary leaves largely open the combination of words into sentences, which is controlled by grammar. And so it goes. Moreover, each lower level imposes restrictions on the one above it, even as the laws of inanimate nature restrict the practicability of conceivable machines; and again, we may observe that a higher operation may fail when the next lower operation escapes from its control. Words are drowned in a flow of random sounds, sentences in a series of random words, and so on.

In a broad way we can see this principle of marginal control operating also in the hierarchy of biotic levels. The vegetative system, sustaining life at rest, leaves open the possibilities of bodily movement by means of muscular action, and the principles of muscular action leave open their integration to innate patterns of behavior. Such patterns leave open, once more, their shaping by intelligence, the working of which offers, in its turn, wide-ranging possibilities for the exercise of still higher principles in those of us who possess them.

These illustrations of the principle of marginality should make it clear that it is present alike in artifacts, like machines; in human performances, like speech; and in living functions at all levels. It underlies the functions of all comprehensive entities having a fixed structure. We may confidently rely, therefore, on our analysis of machines to declare that the predominant view of biologists—that a mechanical explanation of living functions amounts to their explanation in terms of physics and chemistry —is false. Moreover, the conclusion that machines are defined by the fact that boundary conditions expressly left open by physics and chemistry are controlled by principles foreign to physics and chemistry, makes it clear that it is in respect of its characteristic boundary conditions that a mechanically functioning part of life is not explicable in terms of physics and chemistry.

This is not to deny that there is a great deal of truth in the mechanical explanation of life. The organs of the body work much like machines, and they are subject to a hierarchy of controls, exercised by an ascending series of mechanical principles. Biologists pursuing the aim of explaining living functions in terms of machines have achieved astounding success. But this must not obscure the fact that these advances only add to the features of life which cannot be represented in terms of laws manifested in the realm of inanimate nature.

There is an important minority of biologists who deny the possibility of representing all living functions by mechanisms of the kind known to engineering and technology. The non-machinelike processes of life which they postulate, they call *organismic*. Such organismic processes are found at work in regeneration, and are most strikingly demonstrated by the embryonic regeneration of the sea urchin discovered by Hans Driesch. Driesch found that throughout the gastrula stage any cell or combination of cells detached from the embryo will develop into a normal sea urchin. He described an embryo having such regenerative powers as a "harmonious equipotential" system. Such regeneration of the embryo from a fragment is known also as "morphogenetic regulation."

In the course of embryonic development, on the other hand, we find a progressive limitation of equipotentiality due to the fixation of the fate of the several areas of the embryo. This lends the embryo a mosaic character. Two principles are henceforth combined in the development of an embryo. (1) Its division into areas of fixed determination lends it a machinelike structure; (2) the regulative powers which mutually adjust the several areas of fixed potentiality, and which preserve equipotentiality within each area, represent, on the other hand, an organismic principle. As maturation progresses, it leads to increasingly differentiated mechanical structures, and in these the scope of regulation is correspondingly reduced. Biologists who recognize the basic distinction between mechanismic and organismic processes consider living functions to be determined at all stages by a combination of a mechanism with organismic regulation.

Gestalt psychologists have often suggested that the processes of regulation are akin to the shaping of perception, but their insistence that both perceptual shaping and biological regulation are but the result of physical equilibration brought this suggestion to a dead end. I do agree with Gestalt psychologists that the regulative powers of living beings and their mental powers of comprehension are akin to each other, but I believe that they both embody principles that are not manifest in the realm of inanimate nature. Indeed, I regard the presence in living beings of such principles as an established

43

fact, and shall continue to inquire into the operations of these principles.

Let me return to my account of machinelike functions of living beings, based on the principle of marginal control. This control employs distinctive parts of the body, for specific functions. In conjunction with this we now have regulative processes that operate equipotentially. Such operations resemble an integration of particulars by means of tacit knowing, and resemble, above all, the seeing and solving of a problem I have in mind, especially problems like that of composing a poem, inventing a machine, or making a scientific discovery. Such problems are intimations of the potential coherence of hitherto unrelated things and their solution establishes a new comprehensive entity, be it a new poem, a new kind of machine, or a new knowledge of nature.

Inanimate nature is self-contained, achieving nothing, relying on nothing and, hence, unerring. This fact defines the most essential innovation achieved by the emergence of life from the inanimate. A living function has a result which it may achieve or fail to achieve. Processes that are expected to achieve something have a value that is inexplicable in terms of processes having no such value. The logical impossibility of such explanation may be affiliated to the dictum that nothing that *ought* to be, can be determined by knowing what *is*. Hence a principle not present in the inanimate must come into operation when it gives birth to living things.

But the hierarchic structure of the higher forms of life necessitates the assumption of further processes of emergence. If each higher level is to control the boundary conditions left open by the operations of the next lower level, this implies that these boundary conditions are in fact left open by the operations going on at the lower level. In other words, no level can gain control over its own boundary conditions and hence cannot bring into existence a higher level, the operations of which would consist in controlling these boundary conditions. Thus the logical structure of the hierarchy implies that a higher level can come into existence only through a process not manifest in the lower level, a process which thus qualifies as an emergence.

Once more, our understanding of this relationship can be deepened by considering its mental counterpart. The combination of organismic and mechanical principles is replaced in the mental field by the combination of tacit comprehension with a set of fixed logical operations. A child starts off with a scanty repertoire of innate mental connections and enriches them rapidly by using his powers of comprehension for establishing further fixed relations of experience. Piaget has described how a child's powers of reasoning are improved by developing increasingly stable rules of logical procedure. Stimulated by the interiorization of language, this development eventually produces the adult mind. I described in my last lecture the tacit process of comprehension by which we take in the meaning of a communication addressed to us; the process of education by which the human mind is brought into existence is a major exercise of these powers of understanding. The growing mind recreates the whole conceptual framework and all the rules of reasoning bequeathed to it by its culture. In this sense each of these fixations reduces the range of creative powers, but it enlarges these powers, by placing new tools at their disposal. The process works like the anatomical differentiation of a developing organism, which narrows down its areas of equipotentiality, while offering in exchange, the use of a more powerful biotic machinery.

I am accepting here the surmise of Gestalt psychologists, that equipotentialities of the kind discovered by Driesch in the sea urchin embryo are akin to the process by which we shape new ideas. But the kind of emergence that I identify with comprehension is an action which creates new comprehensive entities. It resembles Bergson's *élan vital* and rejects Köhler's dynamic equilibration. We must consider next how this conception applies to the evolution of living beings, for which the presence of such a creative agency was postulated by Bergson, Samuel Butler, and, more recently, Teilhard de Chardin.

We must start by reshaping the problem of evolution deformed by the current theory of evolution. The interest of evolution lies in the rise of higher beings from lower ones and, principally, in the rise of man. A theory which recognizes only evolutionary changes due to the selective advantage of random mutations cannot acknowledge this problem. All forms that continue to survive have the same survival value; only those that are becoming extinct can be said to lack selective advantage. In this respect, man is actually in a bad position today. His survival on earth seems less probable than that of the insects. But this hardly affects the interest we

take in the question of how man came into existence, any more than it makes us turn away from
human history, literature, and arts to the history
and ethology of insects. It is the height of intellectual perversion to renounce, in the name of scientific objectivity, our position as the highest form of
life on earth, and our own advent by a process
of evolution as the most important problem of
evolution.

The misrepresentation of evolution as a process
of continued selective improvement has been as-
sisted by shifting attention from evolution to the
origin of species. A preoccupation with the way
populations of a new kind come into existence has
made us lose sight of the more fundamental question: how any single individual of a higher species
ever came into existence. But we can bring this
problem into focus quite simply, by surveying the
historical antecedents of any single individual of a
higher form.

The origins of one man can be envisaged by tracing the man's family tree all the way back to the
primeval specks of protoplasm in which his first
origins lay. The history of this family tree includes
everything that has contributed to the making of
this human being. This segment of evolution is
precisely on a par with the history of a fertilized
egg developing into a mature man, or the history of
a single plant growing from seed, which includes
everything that caused that man or that plant to
come into existence. Natural selection is concerned
with populations; it plays no part in the evolution
of a single human being. If we insisted on applying
the concepts of population biology to the problem

47

of the individual's origin, this would amount to bringing in possible adversities which did not prevent the line of ancestry from developing, and which of course play no causal part in its development. We might call this more restrictedly causal sequence an *ideogenesis*, in distinction from the usually studied, more comprehensively statistical sequence of *phylogenesis*.*

We do not thereby disregard the occurrence of accidental mutations which may prove adaptive. We merely assume that these can be distinguished from changes of type achieving new levels of existence. Most paleozoologists would agree that, though this distinction is often difficult, it is nonetheless valid. And once this distinction is accepted and allowed for, the autonomous thrust of evolutionary ascent is as clearly manifest as the growth of an individual from a germ cell. To deny it recognition, because scientists have decided that for their professional purposes they should turn a blind eye on it, is to renounce a great truth for the sake of a useful technical fiction.

Throughout my present lecture I have been enlarging on the conception of tacit knowing by means of generalizations which now fall into line with this image of evolution. We had seen our tacit

* Thus the usual chemical kinetics, based on mass action, is a more comprehensively statistical description of chemical change than would be the tracing of the causal chain giving rise to a single new molecule. The modern dynamics of chemical reactions (as studied by D. R. Herschbach, R. B. Bernstein, J. C. Polanyi, and others) is comprised altogether in this causal chain.

powers interpreting the world around us by converting the impacts between our body and the things that come our way into a comprehension of their meaning. This comprehension was both intellectual and practical. The circle of comprehensive entities was then extended to include, apart from our own performances, both the performances of other persons and these persons themselves. Here we reached the door beyond which lay the whole expanse of biology, and the key to the door was supplied by recognizing that comprehensive entities consist in a peculiar logical combination of consecutive levels of reality. Such a conception goes beyond the range of human performances to include all biotic levels within a human being and throughout the realm of animals and plants. It reveals the stratified universe of living things. Within an organism, each higher principle controls the boundary left indeterminate by the next lower principle. It relies for its operations on the lower principle without interfering with its laws, and because the higher principle is logically unaccountable in terms of the lower, it is liable to failure by operating through it.

The first emergence, by which life comes into existence, is the prototype of all subsequent stages of evolution, by which rising forms of life, with their higher principles, emerge into existence. I have included all stages of emergence in an enlarged conception of inventiveness achieved by tacit knowing. The spectacle of rising stages of emergence confirms this generalization by bringing forth at the highest level of evolutionary emergence those mental powers in which we had first recognized our faculty of tacit knowing.

49

My theory of a stratified universe has emphasized the distinctive levels of living things. But evolution is a continuous process: eventually, therefore, my analysis will have to be revised to allow for the gradual intensification of new functions, as they emerge from forms in which they are not yet noticeable. I shall anticipate this here to some extent by speaking of the gradual intensification of a certain group of features through the evolutionary rise of man. The series of increasingly comprehensive operations which lead up to the emergence of man is accompanied at every step by an additional liability to miscarry. The capacity for growth, by which living things acquire their typical shapes, may produce malformations; physiological functions are subject to disabling and eventually mortal diseases; perception, drive satisfaction, and learning bring with them new failings by falling into error; and finally, man is found not only liable to a far greater range of errors than animals are, but, by virtue of his moral sense, becomes capable also of evil.

This parallel development of capabilities and liabilities is accompanied by a consolidation of the center to which these are attributable. Life exists predominantly in the form of individuals. But at the vegetative level, as we have it in plants, individuality is still weak. The center of the individual becomes more pronounced with the rise of animal activities, and it grows more marked still in the exercise of intelligence. It rises to the level of personhood in man. But every additional function with which the center is credited exposes it to new reproaches in respect of new failures.

Since all life is defined by its capacity for success

and failure, all biology is necessarily critical. Observation, strictly free from valuation, is possible only in the sciences of inanimate nature. Traces of criticism are present even in some of these sciences, for example, in crystallography, where it is the specimen which is called imperfect if it does not fit the theory. But crystals don't function; hence the appreciation of even the lowest living being is far more intensely critical than that of a crystal.

Each new branch of biology that was developed to cover the increasingly complex function of higher animals sets up additional standards, to which the observer expects the animal to measure up. And this intensification of criticism coincides with an increasing enrichment of relations between the critic and his object. We know an animal, as we know a person, by entering into its performance, and we appreciate it as an individual, in the interests of which these performances have their meaning. Even at the lowest, purely vegetative level, we accept the interests of the animal as the standard by which our own interest in the animal is determined. All biology is, in this sense, convivial. But this conviviality rises to emotional concern as the animal approaches the human level. We then become aware of its sentience, of its intelligence, and above all of its emotional relations to ourselves.

Yet, however greatly we may love an animal, there is an emotion which no animal can evoke and which is commonly directed toward our fellow men. I have said that at the highest level of personhood we meet man's moral sense, guided by the firmament of his standards. Even when this appears absent, its mere possibility is sufficient to demand our respect.

51

We have here a fact which sets a new major task to the process of evolution: a task which appears the more formidable as we realize that both this moral sense and our respect for it presuppose an obedience to commands accepted in defiance of the immemorial scheme of self-preservation which had dominated the evolutionary process up to this point.

Yet evolution must make sense also of this after-thought to five hundred million years of pure self-seeking. And in a way this problem can be put in biological terms. For this potentiality for obedience to higher demands is largely involved in man's capacity for another peculiarly human relation to other men, namely, the capacity to feel reverence for men greater than oneself. If evolution is to include the rise of man, with all his sense of higher obligations, it must include also the rise of human greatness.

In my last lecture I shall expand the cosmic panorama that I have sketched out so far, to include man's cultural equipment, and this should offer us a framework within which we can define responsible human action, of which man's moral decisions form but a particular instance.

— 3 —

A SOCIETY OF
EXPLORERS

M Y FIRST lecture dealt with our power of tacit knowing. It showed that tacit knowing achieves comprehension by indwelling, and that all knowledge consists of or is rooted in such acts of comprehension. The second lecture showed how the structure of tacit knowing determines the structure of comprehensive entities. By studying the way tacit knowing comprehends human performances, we saw that what is comprehended has the same structure as the act that comprehends it. The relation of a comprehensive entity to its particulars was then seen to be the relation between two levels of reality, the higher one controlling the marginal conditions left indeterminate by the principles governing the lower one. Such levels were then stacked on top of each other to form a hierarchy, and this stacking opened up the panorama of stratified living beings. This stratification offered a framework for defining emergence as the action which produces the next higher level, first from the inanimate to the living and then from each biotic level to the one above it. This holds both for the development of an individual and for the evolution of living things.

Thus emergence took over from tacit knowing the function of producing fundamental innovations; but, as emergence continued to scale the heights leading on to the rise of man, it gradually resumed its encountered form of human knowing. So in the end we were confronted again with the mind of man, making ever new sense of the world by dwelling in its particulars with a view to their comprehension.

Now we enter a new range of subjects. We must ask whether intellectual powers, grounded in tacit knowing and descended from evolutionary emergence, can exercise the kind of responsible judgment which we must claim if we are to attribute a moral sense to man. Could, in fact, my rebuttal of exactitude as the ideal of science open the way toward a theory re-establishing the justification of moral standards?

Let me take my bearings once more from the questions which formed my point of departure. I told you how I was struck by the theory that prevailed for a considerable time under Stalin in Soviet Russia, which denied the justification of science as a pursuit of knowledge for its own sake. I said that this violent self-immolation of the mind was actuated by moral motives, and that a similar fusion of unprecedented critical lucidity and intensified moral passions pervades our whole civilization, inflaming or paralyzing both reason and morality.

The story has often been told how scientific rationalism has impaired moral beliefs, first by shattering their religious sanctions and then by questioning their logical grounds; but the usual account does not explain the state of the modern mind.

It is true that the Enlightenment weakened ecclesiastical authority and that modern positivism has denied justification to all transcendent values. But I do not think that the discredit which the ideal of exact scientific knowledge had cast on the grounds of moral convictions would by itself have much damaged these convictions. The self-destructive tendencies of the modern mind arose only when the influence of scientific skepticism was combined with a fervor that swept modern man in the very opposite direction. Only when a new passion for moral progress was fused with modern scientific skepticism did the typical state of the modern mind emerge.

The new social aspirations had their origins in Christianity, but they were evoked by the attacks on Christianity. It was only when the philosophy of Enlightenment had weakened the intellectual authority of the Christian churches that Christian aspirations spilled over into man's secular thoughts, and vastly intensified our moral demands on society. The shattering of ecclesiastical control may have been morally damaging in the long run, but its early effect was to raise the standards of social morality.

What is more, scientific skepticism smoothly cooperated at first with the new passions for social betterment. By battling against established authority, skepticism cleared the way for political freedom and humanitarian reforms. Throughout the nineteenth century, scientific rationalism inspired social and moral changes that have improved almost every human relationship, both private and public, throughout Western civilization. Indeed, ever since the French Revolution, and up to our own days, scientific rationalism has been a major influence toward intellectual, moral, and social progress.

Where, then, is the fateful conflict between the moral skepticism of science and the unprecedented moral demands of modern man?

Throughout the very period during which they were so beneficently combined, we can trace the rising undercurrent of their joint destructive influence, and this current finally surfaced and eventually became dominant in the last fifty years. Both scientific skepticism and moral perfectionism had for some time been growing more radical and more irreconcilable, and more deeply ingrained in our thought; and eventually they fused into various unions, each of which embodied a dangerous internal contradiction.

These hybrids of skepticism and perfectionism fall into two classes, one personal, the other political.

The first kind of hybrid can be represented by modern existentialism. Scientific detachment, it says, presents us with a world of bare facts. There is nothing there to justify authority or tradition. These facts are there; for the rest, man's choice is unrestricted. You might expect moral perfectionism to be shocked by this teaching. But no, it rejoices in it. For modern existentialism uses moral skepticism to blast the morality of the existing society as artificial, ideological, hypocritical.

Moral skepticism and moral perfectionism thus combine to discredit all explicit expressions of morality. We have, then, moral passions filled with contempt for their own ideals. And once they shun their own ideals, moral passions can express themselves only in anti-moralism. Professions of absolute self-assertion, gratuitous crime and perversity, self-hatred and despair, remain as the only defenses against a searing self-suspicion of bad faith. Modern existentialists recognize the Marquis de Sade as the earliest moralist of this kind. Dostoyevsky's Stavrogin in *The Possessed* is its classic representation in terms of fiction. Its theory was perhaps first outlined by Nietzsche in *The Genealogy of Morals*. Rimbaud's *Une Saison en Enfer* was its first major epiphany. Modern literature is replete with its professions.

The conception of morality established by this movement eliminates the distinction between good and evil, and it is pointless therefore to express opposition to it by moral reprobation.

The unprecedented critical lucidity of modern man is fused here with his equally unprecedented moral demands and produces an angry absolute individualism. But adjacent to this, the same fusion produces political teachings which sanction the total suppression of the individual. Scientific skepticism and moral perfectionism join forces then in a movement denouncing any appeal to moral ideals as futile and dishonest. Its perfectionism demands a total transformation of society; but this utopian project is not allowed to declare itself. It conceals its moral motives by embodying them in a struggle for power, believed to bring about automatically the aims of utopia. It blindly accepts for this belief the scientific testimony of Marxism. Marxism embodies the boundless moral aspirations of modern man in a theory which protects his ideals from skeptical doubt by denying the reality of moral motives in public life. The power of Marxism lies in uniting the two contradictory forces of the modern mind into a single political doctrine. Thus originated a world-embracing idea, in which moral doubt is frenzied by moral fury and moral fury is armed by scientific nihilism.

Bukharin, explaining urbanely, in the spring of 1935, that scientific truth would no longer be pursued for its own sake under socialism, completed the wheel full circle. Embodied in a scientifically sanctioned political power, moral perfectionism had no place left for truth. Bukharin confirmed this three years later when, facing death, he bore false witness against himself. For to tell the truth would have been to condemn the Revolution, which was unthinkable.

It may appear extravagant to hope that these self-destructive forces may be harmonized by reconsidering the way we know things. If I still believe that a reconsideration of knowledge may be effective today, it is because for some time past, a revulsion has been noticeable against the ideas which brought us to our present state. Both inside and outside the Soviet empire, men are getting weary of ideas sprung from a combination of skepticism and perfectionism. It may be worth trying to go back to our foundations and seek to lay them anew, more truly.

I have prepared for this in certain respects. All I have said implied my repudiation of the grounds on which the absolute intellectual self-determination of man was proclaimed by the great philosophic movement engendered by the Enlightenment. For to acknowledge tacit thought as an indispensable element of all knowing and as the ultimate mental power by which all explicit knowledge is endowed with meaning, is to deny the possibility that each succeeding generation, let alone each member of it, should critically test all the teachings in which it is brought up. Statements explicitly derived from identifiable premises can be critically tested by examining their premises and the process of inference which led to them. But if we know a great deal that we *cannot tell*, and if even that which we know and *can* tell is accepted by us as true only in view of its bearing on a reality beyond it, a reality which may yet manifest itself in the future in an indeterminate range of unsuspected results; if indeed we recognize a great discovery, or else a great personality, as *most real*, owing to the wider range of its yet unknown future manifestations: then the idea of knowledge based on wholly identifiable grounds collapses, and we must conclude that the transmission of knowledge from one generation to the other must be predominantly tacit.

We have seen that tacit knowledge dwells in our awareness of particulars while bearing on an entity which the particulars jointly constitute. In order to share this indwelling, the pupil must presume that a teaching which appears meaningless to start with has in fact a meaning which can be discovered by hitting on the same kind of indwelling as the teacher is practicing. Such an effort is based on accepting the teacher's authority.

Think of the amazing deployment of the infant mind. It is spurred by a blaze of confidence, surmising the hidden meaning of speech and adult behavior. This is how it grasps their meaning. And each new step can be achieved only by entrusting oneself to this extent to a teacher or leader. St. Augustine observed this, when he taught: "Unless you believe, you shall not understand."

It appears then that traditionalism, which re-

61

quires us to believe before we know, and in order
that we may know, is based on a deeper insight into
the nature of knowledge and of the communication
of knowledge than is a scientific rationalism that
would permit us to believe only explicit statements
based on tangible data and derived from these by
a formal inference, open to repeated testing.

But I am not reasserting traditionalism for the
purpose of supporting dogma. To argue, as I do,
that confidence in authority is indispensable for the
transmission of any human culture is not to demand submission to religious authority. I admit
that my reaffirmation of traditionalism might have a
bearing on religious thought, but I want to set this
aside here. Modern man's critical incisiveness must
be reconciled with his unlimited moral demands,
first of all, on secular grounds. The enfeebled authority of revealed religion cannot achieve this rec-
onciliation; it may rather hope to be revived by its
achievement.

I will not resist in any way the momentum of the
French Revolution. I accept its dynamism. But I believe that the new self-determination of man can
be saved from destroying itself only by recognizing
its own limits in an authoritative traditional framework which upholds it. Tom Paine could proclaim
the right of each generation to determine its institutions anew, since the range of his demands was
in fact very modest. He unquestioningly accepted
the continuity of culture and of the order of private
property as the framework of self-determination.
Today the ideas of Tom Paine can be saved from
self-destruction only by a conscious reaffirmation
of traditional continuity. Paine's ideal of unlimited

gradual progress can be saved from destruction by revolution only by the kind of traditionalism taught by Paine's opponent, Edmund Burke.

Let me display the inescapable need for a traditional framework first in one example of great modern endeavor, which may then serve as a paradigm for other intellectual and moral progress in a free, dynamic society. My example will be the pursuit of the natural sciences. This may be surprising, for modern science was founded through the violent rejection of authority. Throughout the formative centuries of modern science, the revolt against authority was its battle cry: it was sounded by Bacon and Descartes, and by the founders of the Royal Society in their device, *Nullius in Verba*. What these men said was true and important at the time, but once the adversaries they fought had been defeated, the repudiation of all authority or tradition by science became a misleading slogan.[8]

The popular conception of science teaches that science is a collection of observable facts, which anybody can verify for himself. We have seen that this is not true in the case of expert knowledge, as in diagnosing a disease. But it is not true either in the physical sciences. In the first place, you cannot possibly get hold of the equipment for testing, for example, a statement of astronomy or of chemistry. And supposing you could somehow get the use of an observatory or a chemical laboratory, you would probably damage their instruments beyond repair before you ever made an observation. And even if you should succeed in carrying out an observation to check upon a statement of science and you found a result which contradicted it, you would rightly assume that you had made a mistake.

The acceptance of scientific statements by laymen is based on authority, and this is true to nearly the same extent for scientists using results from branches of science other than their own. Scientists must rely heavily for their facts on the authority of fellow scientists.

This authority is enforced in an even more personal manner in the control exercised by scientists over the channels through which contributions are submitted to all other scientists. Only offerings that are deemed sufficiently plausible are accepted for publication in scientific journals, and what is rejected will be ignored by science. Such decisions are based on fundamental convictions about the nature of things and about the method which is therefore likely to yield results of scientific merit. These beliefs and the art of scientific inquiry based on them are hardly codified: they are, in the main, tacitly implied in the traditional pursuit of scientific inquiry.

To show what I mean I shall recall an example of a claim lacking plausibility to the point of being absurd, which I picked up twenty-five years ago in a letter published by *Nature*. The author of this letter had observed that the average gestation period of different animals ranging from rabbits to cows was an integer multiple of the number π. The evidence he produced was ample, the agreement good. Yet the acceptance of this contribution by the journal was meant only as a joke. No amount of evidence would convince a modern biologist that gestation periods are equal to integer multiples of π. Our conception of the nature of things tells us that such a relationship is absurd, but cannot prescribe how one could prove this. Another, more technical, example from physics can be found in a paper by Lord Rayleigh, published in the *Proceedings of the Royal Society* in 1947. It described some fairly simple experiments which proved, in the author's opinion, that a hydrogen atom impinging on a metal wire could transmit to it energies ranging up to a hundred electron volts. Such an observation, if correct, would be far more revolutionary than the discovery of atomic fission by Otto Hahn in 1939. Yet when this paper appeared and I asked various physicists' opinions about it, they only shrugged their shoulders. They could not find fault with the experiment, yet they not only did not believe its results, but did not even think it worth while to consider what was wrong with it, let alone check up on it. They just ignored it. About ten years later some experiments were brought to my notice which accidentally offered an explanation of Lord Rayleigh's findings. His results were apparently due to some hidden factors of no great interest, but which he could hardly have identified at the time. He should have ignored his observation, for he ought to have known that there must be something wrong with it.[9]

The rejection of implausible claims has often proved mistaken, but safety against this danger could be assured only at the cost of permitting journals to be swamped by nonsense.

There is another requirement to be sustained by the authority of scientific opinion over scientists.

To form part of science, a statement of fact must be not only true, but also interesting, and, more particularly, interesting to science. Reliability, exactitude, does count as a factor contributing to scientific merit, but it is not enough. Two further elements enter into the assessment of a contribution. One is the relation to the systematic structure of science, to correcting or expanding this structure. The other is quite independent of both the reliability and the systematic interest of a discovery, for it lies in its subject matter, in its subject as known before it was taken up by science: it consists of the intrinsic interest of the subject matter.

Thus the scientific interest—or scientific value— of a contribution is formed by three factors: its *exactitude*, its *systematic importance*, and the *intrinsic interest of its subject matter*. The proportion in which these factors enter into scientific value varies greatly over the several domains of science: deficiency in one may be balanced by excellence in another. The highest degree of exactitude and widest range of systematization are found in mathematical physics, and this compensates here for a lesser intrinsic interest of its inanimate subject. By contrast, we have, at the other end of the sciences, domains like zoology and botany which lack exactitude and have no systematic structure comparable in range to that of physics, but which make up for this deficiency by the far greater intrinsic interest in living things compared with inanimate matter. The body of scientific knowledge is what it is by virtue of the fact that referees are constantly engaged in eliminating contributions offered to science which lack an acceptable scientific value, as measured by the compounded coefficients of accuracy, systematic interest, and lay interest of subject matter. Science is shaped by these delicate valuations by scientific opinion.

We may wonder, then, how the conformity enforced by current judgments of plausibility can allow the appearance of any true originality. For it certainly does allow it: science presents a panorama of surprising developments. How can such surprises be produced on effectively dogmatic grounds?

We often hear of *surprising confirmations* of a theory. The discovery of America by Columbus was a surprising confirmation of the earth's sphericity; the discovery of electron diffraction was a surprising confirmation of De Broglie's wave theory of matter; the discoveries of genetics brought surprising confirmations of the Mendelian principles of heredity. We have here the paradigm of all progress in science: discoveries are made by pursuing possibilities suggested by existing knowledge.

This applies also to radically novel discoveries. All the material on which Max Planck founded his quantum theory in 1900 was open to inspection by all other physicists. He alone saw inscribed in it a new order transforming the outlook of man. No other scientist had any inkling of this vision; it was more solitary even than Einstein's discoveries. Although many striking confirmations of it followed within a few years, so strange was Planck's idea that it took eleven years for quantum theory to gain final acceptance by leading physicists. As late as 1914, the controversy over quanta was still sufficiently alive to serve as a joke at a dinner party in the home of the great Walther Nernst in Berlin. A graduate student named Lindemann (who later became Lord Cherwell), said of a fellow student who had just married a rich girl, that he had hitherto been equipartitionist, but now believed in quanta. Yet in another thirty years, Planck's position in science was approaching that hitherto accorded only to Newton.

While science imposes an immense range of authoritative pronouncements, it not merely tolerates dissent in some particulars, but grants its highest encouragement to creative dissent. While the machinery of scientific institutions severely suppresses suggested contributions, because they contradict the currently accepted view about the nature of things, the same scientific authorities pay their highest homage to ideas which sharply modify these accepted views.

This apparent self-contradiction is resolved on the metaphysical grounds which underlie all our knowledge of the external world. The sight of a solid object indicates that it has both another side and a hidden interior, which we could explore; the sight of another person points at unlimited hidden workings of his mind and body. Perception has this inexhaustible profundity, because what we perceive is an aspect of reality, and aspects of reality are clues to boundless undisclosed, and perhaps yet unthinkable, experiences. This is what the existing body of scientific thought means to the productive scientist: he sees in it an aspect of reality which, as such, promises to be an inexhaustible source of new, promising problems. And his work bears this out; science continues to be fruitful—as I said in my first lecture—because it offers an insight into the nature of reality.

This view of science merely recognizes what all scientists believe—that science offers us an aspect of reality, and may therefore manifest its truth inexhaustibly and often surprisingly in the future. Only in this belief can the scientist conceive problems, pursue inquiries, claim discoveries; this belief is the ground on which he teaches his students and exercises his authority over the public. And it is by transmitting this belief to succeeding generations that scientists grant their pupils independent grounds from which to start toward their own discoveries, possibly in opposition to their teachers.

The discovery of new facts may change the interest of established facts, and intellectual standards themselves are subject to change. Interest in spectroscopy was sharply renewed by Bohr's theory of atomic structure, and the novelty of its appeal also wrought a change in the standards of scientific beauty. No single achievement has equaled Planck's discovery of quantum theory in transforming the quality of intellectual satisfaction in mathematical physics. Such changes have been accompanied through the centuries by the belief that they offered a deeper understanding of reality. This testifies to a belief in the reality of scientific value. Only by holding this belief can the scientist direct his inquiries toward tasks promising scientific value. And only in this conviction can he inaugurate novel standards with universal intent. He can then also teach his students to respect current values and encourage them perhaps one day to deepen these values in the light of their own insights.

69

But a description of scientific procedure implies no justification of it. If I trust, as I do, that the metaphysical beliefs of scientists necessarily assure discipline and foster originality in science, I must declare these beliefs to be true. I do this. Yet this does not mean that I share all accepted beliefs of scientists about the nature of things. On the contrary, all my writings show that I dissent from large tracts of scientific views, particularly in psychology and sociology. But this leaves my beliefs about the nature of external scientific truth unaffected.

These metaphysical beliefs are not explicitly professed today by scientists, let alone by the general public. Modern science arose claiming to be grounded in experience and not on a metaphysics derived from first principles. My assertion that science can have discipline and originality only if it believes that the facts and values of science bear on a still unrevealed reality, stands in opposition to the current philosophic conception of scientific knowledge.

❖ ❖ ❖

I have next to answer some curious questions. Research is pursued by thousands of independent scientists all over the planet, each of whom really knows only a tiny part of science. How do the results of these inquiries, each conducted largely in ignorance of the others' work, sustain the systematic unity of science? And how can many thousands of scientists, each of whom has a detailed knowledge of only a very small fraction of science, jointly impose equal standards on the whole range of vastly different sciences?

70

Today's system of science has grown from the system as it was a generation ago through advances which originated independently at a great number of points where the old system offered a chance for progress. Spread over the different fields of scientific knowledge, scientists were looking out for such points and each developed one. Each has studied the work of others on various promising points and also considered for his choice how he could best make use of his special gifts. Such a procedure achieves the greatest total progress possible in practice and best assures the systematic character of science at successive stages of its progress. Such is the work of self-co-ordination by mutual adjustment in science.

All institutions serving the advancement and dissemination of science rely on the supposition that a field of potential systematic progress exists, ready to be revealed by the independent initiative of individual scientists. It is in view of this belief that scientists are appointed for life to the pursuit of research and permanent subsidies are granted to them for this purpose. Many expensive buildings, pieces of equipment, journals, etc. are founded and maintained in this belief. It is the most general traditional belief which a novice joining the scientific community accepts in becoming a scientist.

This raises another, more intricate, question. How can we confidently speak of science as a systematic body of knowledge and assume that it is sufficiently reliable and interesting in all its branches as judged by the same standards of scientific merit? Can we possibly be assured that new contributions will be accepted in all areas by the same standards of plausibility and be rewarded by the same standards of beauty and originality? Unless contributions are accepted in different areas by substantially equal standards, a gross waste of resources might ensue. Can such a scandal be guarded against by transferring resources from areas where standards are currently lower, to points at which they are higher?

It might seem impossible to compare the complex scientific value of marginal contributions over such different areas as, for example, astronomy and medicine. Yet I believe this is in fact done, or at least is reasonably approached in practice. It is done by applying a principle that I have not seen described elsewhere, although it is used in various fields; I would call it the *principle of mutual control*. It consists, in the present case, of the simple fact that scientists keep watch over each other. Each scientist is both subject to criticism by all others and encouraged by their appreciation of him. This is how *scientific opinion* is formed, which enforces scientific standards and regulates the distribution of professional opportunities. It is clear that only fellow scientists working in closely related fields are competent to exercise direct authority over each other; but their personal fields will form *chains of overlapping neighborhoods* extending over the entire range of science. It is enough that the standards of plausibility and worthwhileness be equal around every single point to keep them equal over all the sciences. Even those in the most widely separated branches of science will then rely on each other's results and support each other against any laymen seriously challenging their authority.

Such mutual control produces a mediated consensus between scientists even if they cannot understand more than a vague outline of each other's subjects. And this applies, of course, also to myself. All that I have said here about the workings of mutual adjustment and mutual authority is based on my own personal belief that the modes of intercourse I have observed in my part of science can be assumed to extend through all sciences.*

Mutual control applies also to those newly joining the scientific community at any particular point of its vast domain. They start their inquiries by joining the interplay of mutual co-ordination and at the same time taking their own part in the existing system of mutual control, and they do so in the belief that its current standards are essentially true and common throughout science. They trust the traditions fostered by this system of mutual control without much experience of it and at the same time claim an independent position from which they may reinterpret and possibly revolutionize this tradition.

The scientific community shows some hierarchi-

* I shall explain in a moment how mutual authority also governs other cultural fields. But I think it applies to further consensual activities of which the participants know only a small fragment. It suggests a way by which resources can be rationally distributed between any rival purposes that cannot be valued in terms of money. All cases of public expenditure serving collective interests are of this kind. This is, I believe, how the claims of a thousand government departments can be fairly rationally adjudicated, although no single person can know closely more than a tiny fraction of them.

cal features, but these do not alter the fact that the authority of scientific opinion is exercised by the mutual control of independent scientists, far beyond the direct scope of any one of them. When we speak of science and its progress, or its history, or speak of the standards of science and call them "scientific," we refer to a thing we call "science" of which no one has ever known more than a tiny fragment. We have seen that the tradition of science induces its own renewal by bearing on a reality beyond experience; now we find likewise that each scientist's knowledge of his own neighborhood bears on the whole of science far beyond his own experience. This is how he controls the standards of science indirectly, on the same footing of independence as all others do, while submitting to their control of his own work in return.

Each exchange of mutual criticism is something of a tussle and may be a mortal struggle. New standards of plausibility and of scientific interest are thus initiated and eventually established. The movement will start in one field and gradually diffuse into all others. This is how science is steadily reshaped and yet its coherence maintained over all its fields. I myself am involved in such a predicament today through my dissent from some methods of current psychology and sociology. But I do not challenge thereby the existence of science as a coherent system of thought: I merely press for its reform in certain respects.

✧　✧　✧

We have seen that the scientist can conceive problems and pursue their investigation only by be-

lieving in a hidden reality on which science bears.
Now that I have shown further how scientific originality springs from scientific tradition and at the
same time supersedes it, I can show how this process establishes the sense of personal responsibility
which sustains the scientist's search.

There are two possible ways of viewing the progress made by the front line of scientific discoveries
as it advances over a period of time. We may look
upon such progress as the growth of thought in the
minds of gifted people along the pathways of science. The frequent occurrence of simultaneous discoveries may appear to support this image. Even
major discoveries, which fundamentally refashion
our conception of nature, can be made simultane-
ously by a number of scientists at different places.
Quantum mechanics was discovered in 1925 by
three authors so independent of each other that
they were thought at the time to have given mutually incompatible solutions to the problem. Thus
seen, the growth of new ideas appears altogether
predetermined. The mind of those making discoveries seems merely to offer a suitable soil for the
proliferation of new ideas.

Yet, looking *forward* before the event, the act of
discovery appears personal and indeterminate. It
starts with the solitary intimations of a problem, of
bits and pieces here and there which seem to offer
clues to something hidden. They look like fragments of a yet unknown coherent whole. This ten-
tative vision must turn into a personal obsession;
for a problem that does not worry us is no problem:
there is no drive in it, it does not exist. This obsession, which spurs and guides us, is about something

that no one can tell: its content is undefinable, indeterminate, strictly personal. Indeed, the process by which it will be brought to light will be acknowledged as a discovery precisely because it could not have been achieved by any persistence in applying explicit rules to given facts. The true discoverer will be acclaimed for the daring feat of his imagination, which crossed uncharted seas of possible thought.

Thus the backward-looking picture of thought as using human brains as the passive soil of its proliferation proves false. Yet it does represent an aspect of the pursuit of science. Scientific progress seen after the event may be taken to represent the possibilities that were previously hidden and dimly anticipated in a problem. This does explain how different scientists may independently feel intimations of a particular potentiality, often sighting it by different clues and possibly discovering it in different terms.

Note that there is a widespread opinion that scientists hit on discoveries merely by trying everything as it happens to cross their minds. This opinion follows from an inability to recognize man's capacity for anticipating the approach of hidden truth. The scientist's surmises or hunches are the spurs and pointers of his search. They involve high stakes, as hazardous as their prospects are fascinating. The time and money, the prestige and self-confidence gambled away in disappointing guesses will soon exhaust a scientist's courage and standing. His gropings are weighty decisions.

Such are the responsible choices made in the course of scientific inquiry. The choices are made by the scientist: they are his acts, but what he pursues is not of his making; his acts stand under the judgment of the hidden reality he seeks to uncover. His vision of the problem, his obsession with it, and his final leap to discovery are all filled from beginning to end with an obligation to an external objective. In these intensely personal acts, therefore, there is no self-will. Originality is commanded at every stage by a sense of responsibility for advancing the growth of truth in men's minds. Its freedom is perfect service.

Many writers have observed, since Dewey taught it at the close of the last century, that, to some degree, we shape all knowledge in the way we know it. This appears to leave knowledge open to the whims of the observer. But the pursuit of science has shown us how even in the shaping of his own anticipations the knower is controlled by impersonal requirements. His acts are personal judgments exercised responsibly with a view to a reality with which he is seeking to establish contact. This holds for all seeking and finding of external truth.

There is no more positive justification than this for accepting science to be true. Attempts have been made to compensate for this apparent deficiency by reducing the claims of science. The uncertainty and transiency of science was emphasized and exaggerated for this purpose. Yet this is beside the point. The affirmation of a *probable* statement includes a judgment no less personal than an affirmation of its *certainty* would do. Any conclusion, be it given as a surmise or claimed as a certainty, represents a commitment of the person who arrives at it. No one can utter more than a responsible

77

commitment of his own, and this completely fulfills the finding of the truth and the telling of it.

For the scientist, having relied throughout his inquiry on the presence of something real hidden out there, will justly rely on that external presence also for claiming the validity of the result that satisfies his quest. As he accepted throughout the discipline which the external pole of his endeavor imposed upon him, he expects that others—if similarly equipped—will also recognize the presence that guided him. By his own command, which bound him to the quest of reality, he will claim that his results are universally valid. Such is the universal intent of a scientific discovery.

I speak not of an *established* universality, but of a universal *intent*, for the scientist cannot know whether his claims will be accepted. They may prove false or, though true, may fail to carry conviction. He may even expect that his conclusions will prove unacceptable, and in any case their acceptance will not guarantee him their truth. To claim validity for a statement merely declares that it *ought* to be accepted by all. The affirmation of scientific truth has an obligatory character which it shares with other valuations, declared universal by our own respect for them.

I have spoken of the excitement of problems, of an obsession with hunches and visions that are indispensable spurs and pointers to discovery. But science is supposed to be dispassionate. There is indeed an idealization of this current today, which deems the scientist not only indifferent to the outcome of his surmises, but actually seeking their refutation.[10] This is not only contrary to experience, but logically inconceivable. The surmises of a working scientist are *born of the imagination seeking discovery.* Such effort *risks* defeat but never *seeks* it; it is in fact his craving for success that makes the scientist take the risk of failure. There is no other way. Courts of law employ two separate lawyers to argue opposite pleas, because it is only by a passionate commitment to a particular view that the imagination can discover the evidence that supports it.

The creative thrust of the imagination is fed by various sources. The beauty of the anticipated discovery and the excitement of its solitary achievement contribute to it in the first place. The scientist also seeks professional success, and, if scientific opinion rewards merit rightly, ambition too will serve as a true spur to discovery.

The part of science for the renewal of which the scientist assumes responsibility is surrounded by a sea of information on which he must rely for his enterprise. The scientist may regard his selected field as his "calling," which necessarily includes his submission to the vast area of information and belief surrounding his selected field of inquiry. Each scientist's calling has a different geography. Each must try to choose a problem that is not larger or more difficult than he can master. His faculties would not be fully utilized if he applied them to a lesser task, and would be altogether wasted on a larger one. The degree of originality any particular scientist trusts himself to possess should thus determine the range which he will venture to tackle and hence also the range of information which he will unquestioningly accept. Goethe

79

wrote that the master proves himself by his restraint—and the same holds for science.

My account of scientific discovery describes an existential choice. We start the pursuit of discovery by pouring ourselves into the subsidiary elements of a problem and we continue to spill ourselves into further clues as we advance further, so that we arrive at discovery fully committed to it as an aspect of reality. These choices create in us a new existence, which challenges others to transform themselves in its image. To this extent, then, "existence precedes essence," that is, it comes before the truth that we establish and make our own.

But does this show us that "man is his own beginning, author of all his values"? If originality in science is taken as an example of existential choice, these claims of Nietzsche's and Sartre's existentialism appear ill-conceived. The most daring innovations of science spring from a vast range of information which the scientist accepts unchallenged as a background to his problem. Even when he is led to modify the standards of scientific merit, current standards will be the basis of this reform. Science as a whole, as mediated by thousands of fellow scientists he has never heard about, he accepts unchallenged.

His quest transforms him by compelling him to make a sequence of choices. Does this mean that he is existentially choosing himself? In a sense it does; he does seek intellectual growth. But he does *not* sit back and choose at his pleasure a new existence. He strains his imagination to the utmost to find a path that might lead to a superior life of the mind. All his existential choices are made in response to a potential discovery; they consist in sensing and following a gradient of understanding which will lead to the expansion of his mental existence. Every step is an effort to meet an immediate necessity; his freedom is continuous service.

There is here no existential choice comprising the whole world and claiming responsibility for it. Such a choice would leave neither a center to which it could be responsible, nor a criterion by which it could be judged. This impossible responsibility, which is the source of the existentialist's sense of universal absurdity, now appears as an obvious self-contradiction.

At the opposite extreme, the theory of science taught under Stalin is equally erroneous. To declare that in a classless society the pursuit of science—the search marked by the names of Copernicus and Newton, of Harvey, Darwin, and Einstein—will spontaneously turn to the advancement of the next Five-Year Plan is simply ludicrous.

Yet this doctrine was not without grave consequences in Soviet Russia and not without influence in England. It was my attempt to vindicate the freedom of science against such teachings that made me realize the weakness of the strict empiricism which has dominated our conception of science throughout this century. I saw that this philosophy left science defenseless against the Soviet doctrine and this led me to decide that only on metaphysical grounds can we account for the intrinsic powers of human inventiveness. Here I met also the presuppositions of freedom in science.

81

But more insistent than the imposition of dialectical materialism on science was the subjection of literature and the arts to socialist realism, and of the very conception of truth, morality, and justice to partyism. These doctrines gained force around 1932 and have been substantially relaxed only since the Hungarian and Polish revolutions of 1956. They were to embody man's homeless ideals in an absolute submission to the Communist party.

I have countered the attempt at transforming science on such lines by offering solid grounds for its independence. This would now have to be expanded to all other major principles of man. I cannot attempt this here, but I shall outline—however sketchily—those foundations of science which broadly hold also for all other creative systems of the modern mind.

Scientific tradition derives its capacity for self-renewal from its belief in the presence of a hidden reality, of which current science is one aspect, while other aspects of it are to be revealed by future discoveries. Any tradition fostering the progress of thought must have this intention: to teach its current ideas as stages leading on to unknown truths which, when discovered, might dissent from the very teachings which engendered them. Such a tradition assures the independence of its followers by transmitting the conviction that thought has intrinsic powers, to be evoked in men's minds by intimations of hidden truths. It respects the individual for being capable of such response: for being able to see a problem not visible to others, and to explore it on his own responsibility. Such are the metaphysical grounds of intellectual life in a free, dynamic

82

society: the principles which safeguard intellectual life in such a society. I call this a society of explorers.

In a society of explorers man is *in thought*. Man the explorer is placed in the midst of potential discoveries, which offer him the possibility of numberless problems. We have seen how scientists, scattered over the globe, respond to one vast field of potential thought, how each finds in it a congenial area to develop, and how the results then coordinate themselves to produce a systematic expansion of science. This is also how other kinds of thought have developed in our time. Our age has seen an unprecedented wealth of literary and artistic movements growing into coherence. The ideas of the Enlightenment bred scientism and romanticism in a multitude of connected forms; and since the turn of the eighteenth century legal and social reforms have humanized life in a hundred mutually related ways. This is also how the hybrid of absolute skepticism and perfectionism has engendered in the present century new movements of fiction, poetry, music, and painting—although admittedly this movement has brought forth at the same time the theories prefiguring modern fanaticisms, with all their tyrannies and cruelties.

The structure of authority exercised over a society of explorers is different from that to which a dogmatic society submits. Take once more the example of science. I have spoken of the principle of mutual control through which each scientist independently plays his part in maintaining scientific traditions over an immense domain of inquiry of which he knows virtually nothing. A society of explorers is controlled throughout by such mutually imposed authority. The pressure exercised by literary and artistic circles is notorious. They control access to public recognition much as scientific opinion controls it for science. Their professional opinion commands lay assent just as scientific opinion does. There are, of course, differences: reliance on secondhand authority reaches less far in literature and art than in science, and divisions between rival opinions go deeper. In our society, ideas about morality are also actively cultivated by different circles of mutual appreciation, which are deeply divided against each other; and in politics these circles are deliberately organized as rivals.

But we need not go into all these variations; they are transcended by a test which proves that all such groups effectively foster the intrinsic power of thought. For these circles, these professional associations—some perhaps no more than coteries of mutual admiration—are feared and hated by modern totalitarian rulers. They are feared because in them man lives in thought—in thought over which the rulers have no power. They are feared more than are scientific associations, because the truth of literature and poetry, of history and political thought, of philosophy, morality, and legal principles, is more vital than the truth of science. This is why the independent cultivation of such truth has proved an intolerable menace to modern tyranny.

I have now roughly generalized the principles underlying the pursuit of science to include the cultivation of man's other ideals. The result shows how closely the growth of thought intrinsically limits our self-determination everywhere. Whether his calling lies in literature or art, or in moral and social re-

84

form, even the most revolutionary mind must choose as his calling a small area of responsibility, for the transformation of which he will rely on the surrounding world as its premises. Perfectionism, which would transform the whole of thought and the entire society, is a program of destruction, ending up at best in a world of pretense. The existentialist contempt for all values not chosen by ourselves, condemning them as bad faith, is likewise either empty or destructive.

There is another way of dealing with the claim to absolute self-determination and the demands of perfectionism. I could reject these inordinate endeavors by referring to the logic by which successive levels of reality are related to each other. All our higher principles must rely for their working on a lower level of reality and this necessarily sets limits to their scope, yet does not make them reducible to the terms of the lower level. This argument confutes the current cultural movement which questions the intrinsic powers of our ideals. There is not one higher principle of our minds that is not in danger of being falsely explained away by psychological or sociological analysis, by historical determinism, by mechanical models or computers; but this battle cannot be joined here on this wide front.

Nevertheless, I must say a word on these lines about the principal ideal of man which is at the core of his involvement in a combination of extreme skepticism and perfectionism, for I have specifically promised to find a place for moral principles safe from self-destruction by a claim to boundless self-determination. Take the demand for social and moral perfection and recognize the presence of successive levels of reality. Society, as an organization of power and profit, forms one level, while its moral principles lie on a level above it. The higher level is rooted in the lower one: moral progress can be achieved only within the medium of a society operating by the exercise of power and aiming at material advantages. We must accept the fact that any moral advances must be tainted by this social mechanism which alone can bring them about. To attempt to enforce absolute morality in society is therefore to indulge in fantasies that will only lead to untamed violence.

The problem of a balanced mind, secured against both critical and moral frenzy, has gained new and novel urgency through the liberating movement that has arisen in the Soviet empire since Stalin's death, a movement which gradually widened into the rebellion of leading young Communists in Hungary, in Poland, and eventually in Russia itself. The question now is whether this revulsion can generate a steady movement of free thought. When I listen to my Hungarian friends who took refuge in England after taking part in the revolution of 1956, when I read their account of their times as ardent Stalinists and of the change of heart they have undergone since then, I find that their hopes are basically the same as those which animated liberal thought at the turn of the last century. They are the hopes with which I was brought up as a child in Hungary. But the innocence with which I breathed in these ideals cannot be recovered. The revival of the liberal tradition can be assured only if we can establish it on a new, conscious understanding of its foundation, on grounds which will withstand modern self-doubt coupled with perfectionism.

With this in mind, let me add a few more touches to the cosmic background of the ideas which I have tried to develop in answer to the need of our time.

I have shown how man can exercise responsible judgment when faced with a problem. His decisions in casting around for a solution are necessarily indeterminate, in the sense that the solution of an unsolved problem is indeterminate; but his decisions are also responsible in being subject to the obligation to seek the predetermined solution of his problem. I have said that this is a commitment to the anticipation of a hidden reality, a commitment of the same kind as exemplified in the knowledge of scientific truth. Responsibility and truth are in fact but two aspects of such a commitment: the act of judgment is its personal pole and the independent reality on which it bears is its external pole.

Since a problem can be known only tacitly, our knowledge of it can be recognized as valid only by accepting the validity of tacit knowing; and the same applies to truth in its bearing on reality. Herein lies the importance of establishing the validity of tacit knowing. But we must yet consider the evolutionary antecedents of man's power to see a problem and solve it. I have identified the antecedents of problem-solving with the process of emergence. Therefore, if responsible human decisions are incidental to problem-solving, we must expect to find that similar decisive powers are intrinsic to the process of emergence throughout the

87

evolutionary innovations which we ascribe to emergence.

Such speculations are necessarily hazardous, but I feel more repelled by our timidity which would leave these matters to the scientists, who actually regard them as unsuited for investigation by science.

I have said that the laws of nature that are manifest in the inanimate domain fail in two respects to account for the rise of living beings. (1) They leave open the boundary conditions which in living beings are fixed according to operational principles not manifested in inanimate nature, and (2) they contain no reference to sentience, which is a characteristic condition of life in higher animals. It seems reasonable to assume that these two deficiencies are but aspects of a single principle that must be added to those of inanimate nature in order to account for living things. I have, of course, introduced this assumption already by identifying emergence with tacit knowing.

A closer definition of the missing principle can be attempted by considering how it could best form a transition from physics to the process governing the growth of thought in the mind of man. Inanimate nature is controlled by forces which draw matter toward stabler configurations. This is equally true in mechanics and thermodynamics and it applies also to open systems, like flames or flows. The forces generated by stabler potentialities may be held in check by various kinds of friction, which may be overcome by catalytic releasing agents. An explosion can be triggered by the spontaneous disintegration of a single molecule. Quantum mechanics has also established the conception of uncaused

causes, subject only to control by a field of proba-
bilities. The decomposition of a radioactive atom
may be an uncaused cause. Let us then carry forward these three characteristics of inanimate processes. (1) We see forces driving toward stabler
potentialities; (2) catalysts or accidental releasers
of friction-locked forces cause them to actualize
these potentialities; and (3) such accidents may be
uncaused events, subject only to probable tendencies.

Look now at the way innovations are achieved by
the effort of human thought. This process too can
be described as the actualization of certain potentialities. To see a problem and undertake its pursuit
is to see a range of potentialities, believed to be
accessible. Such heuristic tension appears to be
generated in the alert mind, much as forces in physics are generated by the accessibility of stabler configurations. But this tension appears to be deliber-
ate: it is a response striving to comprehend a
solution believed to be predetermined. It makes
choices that are hazardous but always controlled by
the pursuit of their intention. These choices resemble quantum mechanical events in being uncaused
and at the same time guided by a field that leaves
them largely indeterminate. But discoveries differ
from inanimate events in three ways: (1) the field
evoking and guiding them is not that of a more
stable configuration but of a problem; (2) their occurrence is not spontaneous but due to an effort
toward the actualization of certain hidden potentialities; and (3) the uncaused action which evokes
them is usually an imaginative thrust toward discovering these potentialities.

So far there is not much speculation involved in this analysis, and I believe there is also some firm support for generalizing it to the process of evolutionary innovations. My analysis of consecutive operational levels necessitates the assumption of a principle which works in the manner of an innovation achieved by tacit integration. The assumption that this process is evoked by the accessibility of the higher levels of stable meaning which it eventually achieves, seems compelling to me. The tension generated by such a higher potentiality might then be triggered into action either by accident or by the operation of first causes. It seems, furthermore, consonant with the conceptual framework of quantum mechanics on the one hand and of problem-solving on the other hand to assume that these creative releases are controlled, and yet never fully determined, by their potentialities. They may succeed or fail. And it seems reasonable to assume, then, on grounds of continuity, that this peculiar kind of indeterminacy is accompanied by the rise of consciousness.

The overall title I gave to these lectures was Man in Thought. I wanted to speak of the logical interrelation between living and thinking in man and to extend this interrelation by tracing the joint ancestry of man and thought, all the way back to their inanimate antecedents. I have introduced a large number of new principles, crowding them on top of each other. I could not resist the temptation of giving you a glance at ideas that have filled volumes, and perhaps may fill others. But I feel that, actually, all I have spoken of presents a single, fairly simple vision. This part of the universe, in which man has arisen, seems to be filled with a field of potentialities which evoke action. The action thus evoked in inanimate matter is rather poor, perhaps quite meaningless. But dead matter, matter that is both lifeless and deathless, takes on meaning by originating living things. With them a hazard enters the hitherto unerring universe: a hazard of life and death.

The field of new potential meanings was so rich that this enterprise, once started, swept on toward an infinite range of higher meanings, unceasingly pouring them into existence, for the better part of a billion years. Almost from the start, this evolutionary response to potential meaning had its counterpart in the behavior of the living things it brought forth. It seems that even protozoa have the faculty of learning; they respond to potential meaning. Rising stages of evolution produce more meaningful organisms, capable of ever more complex acts of understanding. In the last few thousand years human beings have enormously increased the range of comprehension by equipping our tacit powers with a cultural machinery of language and writing. Immersed in this cultural milieu, we now respond to a much increased range of potential thought.

It is the image of humanity immersed in potential thought that I find revealing for the problems of our day. It rids us of the absurdity of absolute self-determination, yet offers each of us the chance of creative originality, within the fragmentary area which circumscribes our calling. It provides us with the metaphysical grounds and the organizing principle of a Society of Explorers.

Yet the question remains whether this solution

will satisfy us. Can we recognize the limitations it imposes on us? Must not such a fragmented society appear adrift, irresponsible, selfish, apparently chaotic? I have praised the freedom of a community where coherence is spontaneously established by self-co-ordination, authority is exercised by equals over each other, all tasks are set by each to himself. But where are they all going? Nobody knows; they are just piling up works soon to be forgotten.

I have tried to affiliate our creative endeavors to the organic evolution from which we have arisen. This cosmic emergence of meaning is inspiring. But its products were mainly plants and animals that could be satisfied with a brief existence. Men need a purpose which bears on eternity. Truth does that; our ideals do it; and this might be enough, if we could ever be satisfied with our manifest moral shortcomings and with a society which has such shortcomings fatally involved in its workings.

Perhaps this problem cannot be resolved on secular grounds alone. But its religious solution should become more feasible once religious faith is released from pressure by an absurd vision of the universe, and so there will open up instead a meaningful world which could resound to religion.

NOTES

NOTES

(1) Lazarus, R. S., and McCleary, R. A., *Journal of Personality* (Vol. 18, 1949), p. 191, and *Psychological Review* (Vol. 58, 1951), p. 113. These results were called in question by Eriksen, C. W., *Psychological Review* (Vol. 63, 1956), p. 74 and defended by Lazarus, *Psychological Review* (Vol. 63, 1956), p. 343. But in a later paper surveying the whole field—*Psychological Review* (Vol. 67, 1960), p. 279—Eriksen confirmed the experiments of Lazarus and McCleary, and accepted them as evidence of subception.

I am relying on subception only as a confirmation of tacit knowing in an elementary form, capable of quantitative experimental demonstration. For me it is the mechanism underlying the formation of Gestalt, from which I first derived my conception of tacit knowing in *Personal Knowledge*. Strangely enough, the connection of subception with Gestalt has been hardly noticed by psychologists in the course of their controversies on the validity of subception. I could find only one place alluding to it, in a paper by Klein, George S., "On Subliminal Activation," *Journal of Nervous Mental Disorders* (Vol. 128, 1959), pp. 293–301. He observes: "It requires no experimental demonstration to say confidently that we are not aware of all the stimuli which we use in behavior."

I have said already basically in *Personal Knowledge* and have continued to emphasize since then, that it is a mistake to identify subsidiary awareness with unconscious or preconscious awareness, or with the Jamesian fringe of awareness. What makes an awareness subsidiary is the *function it fulfills*; it can have any degree of consciousness, so long as it func-

95

tions as a clue to the object of our focal attention. Klein supports this by saying that subliminal activation is but a special case of *transient or incidental stimuli* of all kinds. It is not the subliminal status that matters but "the meanings and properties [a stimulus] acquires at the periphery of thought and action."

Eriksen and Kuethe, whose observation of not consciously identified avoidance I have quoted as a kind of subception, have called this avoidance a defense mechanism, thus affiliating it to Freudian conceptions. This practice is widespread and has caused *Psychological Abstracts* to divide the subject matter into subception and defense mechanism.

Yet another fragmentation of this matter occurred by taking due notice of Otto Pötzl's observations going back to 1917. A survey of his work and of that of his direct successors has appeared in *Psychological Issues* (Vol. II, No. 3, 1960) under the title "Preconscious Stimulation in Dreams, Associations, and Images" by Otto Pötzl, Rudolf Allers, and Jacob Teler, International Universities Press, New York 11, N.Y. An introduction to this monograph by Charles Fisher links these observations to recent studies and notes the present uncertainty about the status of stimuli of which we become conscious only in terms of their contribution to subsequent experience. "The matter needs to be settled," writes Fisher on p. 33, "because the issue of subliminality has important implications for theories of perception."

I believe that this matter has actually much wider implications and must be generally subsumed under the logical categories of tacit knowing.

(2) Eriksen, C. W., and Kuethe, J. L., "Avoidance Conditioning of Verbal Behavior Without Awareness: A Paradigm of Repression," *Journal of Abnormal and Social Psychology* (Vol. 53, 1956), pp. 203–09.

(3) Hefferline, Ralph F., Keenan, Brian, and Harford, Richard A., "Escape and Avoidance Conditioning in Human Subjects Without Their Observation of the Response," *Science* (Vol. 130, November 1959), pp. 1338–39. Hefferline, Ralph F., and Keenan, Brian, "Amplitude-Induction Gradient of a Small Human Operant in an Escape-Avoidance Situation," *Journal of the Experimental Analysis of Behavior* (Vol. 4, January 1961), pp. 41–43. Hefferline, Ralph F., and Perera, Thomas B., "Proprioceptive Discrimination of a Covert Operant Without Its Observation by the Subject," *Science* (Vol. 139, March 1963), pp. 834–35. Hefferline, Ralph F., and Keenan, Brian, "Amplitude-Induction Gradient of a Small Scale (Covert) Operant," *Journal of the Experimental Analysis of Behavior* (Vol. 6, July 1963), pp. 307–15. See also general conclusions in Hefferline, Ralph F., "Learning Theory and Clinical Psychology—An Eventual Symbiosis?" from *Experimental Foundations of Clinical Psychology*, ed. Arthur J. Bachrach (1962).

Note also that numerous Russian observations, reported by Razran, G., "The Observable Unconscious and the Inferable Conscious," *Psychological Review* (Vol. 68, 1961), p. 81, have established the conditioning of intestinal stimuli, having a similar covert character as Hefferline's muscular twitches.

(4) Dilthey, W., *Gesammelte Schriften* (Vol. VII, Leipzig and Berlin, 1914–36), pp. 213–16; [Translation by H. A. Hodges, *Wilhelm Dilthey* (New York, Oxford University Press, 1944), pp. 121–24].

(5) Lipps, T., *Asthetik* (Hamburg, 1903).

(6) W. Haas, writing as a linguist, has come to similar conclusions about the hierarchic structure of language. See W. Haas, "Relevance in Phonetic Analysis," *Word* (Vol. 15, 1959), pp. 1–18; "Linguistic

Structures," *Word* (Vol. 16, 1960), pp. 251–76; also "Why Linguistics Is Not a Physical Science," paper presented to the International Congress for Logic, Methodology and Philosophy of Science, Jerusalem, Israel, August–September 1964.

(7) See *Personal Knowledge*, pp. 328–35; *Study of Man*, pp. 47–52; "Tacit Knowing and Its Bearing on Some Problems of Philosophy," 1962; "Science and Man's Place in the Universe," 1964; "On the Modern Mind," 1965; "The Structure of Consciousness," 1965; "The Logic of Tacit Inference," 1966. *See* Related Bibliography.

(8) My ideas about the traditional grounds of science, the organization of scientific pursuits, and the cultivation of originality go back to *Science, Faith and Society*. They were partly developed earlier in the essays published in *The Logic of Liberty* and they formed later the basis for *Personal Knowledge*. More recent statements on this are in "Science: Academic and Industrial," 1961, "The Republic of Science," 1962; "The Potential Theory of Adsorption," 1963; "The Growth of Science in Society," 1966. *See* Related Bibliography.

(9) Further particulars on the letter to *Nature* are to be found in *The Logic of Liberty*, p. 17. For Lord Rayleigh, see *The Logic of Liberty*, p. 12 and for the later development of Lord Rayleigh's story, see *Personal Knowledge*, p. 276.

(10) This view has been persuasively expressed by K. R. Popper, e.g. in the *Logic of Scientific Discovery* (New York, 1959), p. 279, as follows:
"But these marvellously imaginative and bold conjectures or 'anticipations' of ours are carefully and soberly controlled by systematic tests. Our method of research is not to defend them, in order to prove how right we were. On the contrary, we try to overthrow

them. Using all the weapons of our logical, mathematical, and technical armoury we try to prove that our anticipations were false—in order to put forward, in their stead, new unjustified and unjustifiable anticipations, new 'rash and premature prejudices', as Bacon derisively called them."

RELATED BIBLIOGRAPHY

Papers by the author on which this book has drawn or in which its ideas are developed further:

"Tyranny and Freedom, Ancient and Modern," *Quest* (Calcutta, 1958).

"The Two Cultures," *Encounter* (September 1959).

"Beyond Nihilism," Eddington Lecture (Cambridge University, 1960); also *Encounter* (1960).

"Faith and Reason," *Journal of Religion* (Vol. 41, 1961), pp. 237–41.

"Knowing and Being," *Mind* (Vol. 70, 1961), pp. 458–70.

"The Study of Man," *Quest* (Calcutta, April–June 1961).

"Science: Academic and Industrial," *Journal of the Institute of Metals* (Vol. 89, 1961), pp. 401–06.

"Clues to an Understanding of Mind and Body," in I. J. Good, ed., *The Scientist Speculates* (Heinemann, 1962).

"History and Hope: An Analysis of Our Age," *Virginia Quarterly Review* (Vol. 38, 1962), pp. 177–95.

"The Republic of Science, Its Political and Economic Theory," *Minerva* (Vol. 1, 1962), pp. 54–73.

"The Unaccountable Element in Science," *Philosophy* (Vol. 37, 1962), pp. 1–14.

"Tacit Knowing and Its Bearing on Some Problems of Philosophy," *Review of Modern Physics* (Vol. 34, 1962), p. 601 ff.

"The Potential Theory of Adsorption: Authority in Science Has Its Uses and Its Dangers," *Science* (Vol. 141, 1963), pp. 1010–13.

"Science and Man's Place in the Universe," in Harry

Woolf, ed., *Science as a Cultural Force* (Johns Hopkins, 1964).

"On the Modern Mind," *Encounter* (May 1965).

"The Structure of Consciousness," *Brain* (Vol. 88, Part IV, 1965), pp. 799–810.

"The Logic of Tacit Inference," *Philosophy* (Vol. 40, 1966), pp. 369–86.

"The Creative Imagination," *Chemical and Engineering News* (Vol. 44, No. 17, 1966).

"The Growth of Science in Society," *Encounter* (1966).

The author's books to which reference is made:

Science, Faith and Society
University of Chicago Press, 1946; Oxford University Press, 1946; Phoenix edition, Chicago, 1964.

The Logic of Liberty
University of Chicago Press, 1951; London, Routledge, 1951.

Personal Knowledge
University of Chicago Press, 1958; London, Routledge, 1958; New York, Harper Torchbooks, 1964.

The Study of Man
University of Chicago Press, 1959; London, Routledge, 1959; Phoenix edition, Chicago, 1964.